Rhetoric and Public Memory in the Science of Disaster

LEXINGTON STUDIES IN CONTEMPORARY RHETORIC

Series Editor: Andrew F. Wood, San Jose State University

This series provides thought-provoking and accessible analyses of the uses of language and media from the middle of the twentieth century to the present. In particular, this series examines how modern discourse is constructed and communicated in our distracted times, focusing on specific settings such as health communication, crisis communication, changing norms of interpersonal exchange, political communication beyond the presidency, etiquettes of communication in the digital age, celebrity as a rhetorical form, changing norms in "reality" and narrative television, and altered patterns of address in computer-mediated discussion. These books will provide depth and clarity about discourse more familiar than understood.

Recent Titles in This Series

Rhetoric and Public Memory in the Science of Disaster
By Jeremy R. Grossman
Nikki Haley's Lessons from the New South
By Wanda Little Fenimore
Rhetoric of the Opioid Epidemic: Deaths of Despair in America
By Tiara Good
A Rhetoric of Ruins: Exploring Landscapes of Abandoned Modernity
By Andrew F. Wood
The Corruption of Ethos in Fortress America: Billionaires, Bodycams, and Bureaucrats
By Christopher Carter
Rhetoric and Governance under Trump: Proclamations from the Bullshit Pulpit
By Bernd Kaussler, Lars J. Kristiansen, and Jeffrey Delbert
Rhetorics Haunting the National Mall: Displaced and Ephemeral Public Memories
Edited by Roger C. Aden
Rethinking Communication in Social Business: How Re-Modeling Communication Keeps Companies Social and Entrepreneurial
By Craig E. Mattson
World War I, Mass Death, and the Birth of the Modern US Soldier: A Rhetorical History
By David W. Seitz
US Public Memory, Rhetoric, and the National Mall
Edited by Roger C. Aden

Rhetoric and Public Memory in the Science of Disaster

Jeremy R. Grossman

LEXINGTON BOOKS
Lanham • Boulder • New York • London

Published by Lexington Books
An imprint of The Rowman & Littlefield Publishing Group, Inc.
4501 Forbes Boulevard, Suite 200, Lanham, Maryland 20706
www.rowman.com

86-90 Paul Street, London EC2A 4NE

British Library Cataloguing in Publication Information Available

Library of Congress Cataloging-in-Publication Data

Library of Congress Cataloging-in-Publication Data
Names: Grossman, Jeremy R., author.
Title: Rhetoric and public memory in the science of disaster / Jeremy R. Grossman.
Description: Lanham : Lexington Books, 2024. | Series: Lexington studies in contemporary rhetoric | Includes bibliographical references. | Summary: "This book grapples with the role of science in the public memory of natural disasters and explores how we remember natural disasters by analyzing how we try to prevent them"—Provided by publisher.
Identifiers: LCCN 2024007863 (print) | LCCN 2024007864 (ebook) | ISBN 9781666938937 (cloth) | ISBN 9781666938944 (ebook)
Subjects: LCSH: Natural disasters—United States—Historiography—Case studies. | Natural disasters—Simulation methods—Social aspects—United States—Case studies. | Technical writing—Social aspects—United States—Case studies. | Collective memory—United States—Case studies. | Language and history—United States—Case studies.
Classification: LCC GB5010 .G76 2024 (print) | LCC GB5010 (ebook) | DDC 303.48/50973—dc23/eng/20240411
LC record available at https://lccn.loc.gov/2024007863
LC ebook record available at https://lccn.loc.gov/2024007864

Acknowledgements

This book transformed considerably over the course of four years, particularly as it became clear that I would not be pursuing a traditional academic career. On the one hand, that fact provided space in my head for the work to be closer to what I envisioned for myself, what I believed to be the project necessary for the conversation rather than the most frictionless one; on the other hand, the lack of a structural motive for carrying out the task made it that much more evident to me how indebted I am to many, many people for helping push the project from conception to completion.

Although an almost entirely different project, my PhD dissertation was the necessary first step in laying the groundwork for what this book ended up being. The most profound thanks go to my advisor, Barbara Biesecker, for always seeing both the more and the less in what I write. Her intellectual vision is matched only by her ability to bestow it onto others. My committee members Roger Stahl, Kelly Happe, and Bradford Vivian enriched and challenged the first iteration in ways that clearly reverberated through to the project's end. Technically, theoretically, and professionally, I would not be the writer I am today without their collective guidance.

To my colleagues who weighed in, read drafts, responded to presentations, and otherwise kept me on the rails over the course of this project, I cannot thank you enough for helping me find its core. This list includes Atilla Hallsby, Jason Myres, Kristy Maddux, Lee Pierce, Emily Winderman, Timothy Barney, and Matthew Salzano.

Chapter 4 was first published in *Rhetoric & Public Affairs* under the late Martin Medhurst's editorship. His guidance and facilitation during the refinement and publication of that essay represents now, for me, anyway, a clear fulcrum in the maturation arc of the project from idea to real scholarly contribution. Many heartfelt thanks to him, Laurel Medhurst, Fielding Montgomery, and the anonymous reviewers.

Genuine gratitude goes to my editor, Nicollette Amstutz, for the initial vote of confidence, constant advocacy, and general administrative deftness.

Thanks also to the staff at the National Archives and Library of Congress. It is a real privilege to live so close to such an incredible group of researchers.

Thanks also go to my people: my parents, each of whom endured the evolution of my thoughts about my writing and its purpose, and my sister, who was always interested despite my occasional lack of clarity. To the animals—Zeppelin, Ellie, Ed, and Dr. Frankenstein.

And finally, to Aya, who models drive, care, and discipline every single day of our lives.

Contents

Introduction

The seeds of this project, begun during my PhD work, posed immediately two distinct challenges. The first involved both the excruciatingly laborious and academically ordinary task of solving the initial major line of inquiry that the project posed, its *raison d'être*, the premise that must be sedimented before one can know how to proceed: why was the public memorialization of Katrina so different from other disasters? The official Hurricane Katrina Memorial in New Orleans "arrived" two years after the storm, but lax construction efforts delayed it another year. There was no national outrage over the forestallment of this symbolic gesture, this exclamation of penance and grief, nor was there outrage over the fact that because its design was to include the bodies of the unclaimed and unidentified, these victims occupied the morally outrageous position of the forgotten, a ridiculous incorporation within the epideictic *milieu*. It was one thing for the city to drag its feet on the infrastructure of mourning of and celebration for all victims of the storm; it should have been quite another to leave suspended in limbo the very bodies whose memorialization could only be achieved by these means, whose flesh relied on the broad memorial gesture of the disaster monument in the absence of a proper signifier to which each body could be attached. But not only the city dragged its feet. The lack of outrage and debate—over the timeframe, over the design, over any symbolic element of the memorial—marked the first symptom to be interpreted, the first question to be answered. What was different about Katrina?

The second challenge followed, logically and chronologically, from the first. Peers, friends, family, and strangers mostly seemed to have an exceedingly simple answer: "Well, Hurricane Katrina was a natural disaster. We don't really memorialize natural disasters." This is manifestly true. My current home, Washington, D.C., is an urban geography that rivals Europe in its commitment to garish monumentality, and the marble and bronze structures that I pass every day rarely serve as memory tributes to the victims of natural disasters. However, that we don't really memorialize natural disasters is an observation, not an explanation. It is also an impossible observation to square with an increasingly robust base of literature that argues, in effect, that there is no such thing as a natural disaster. Once the concept of disaster transformed with the rise of the modern era from a metaphysical affliction into a problem of people, nature, and governments, its conceptual status and its political and material elaborations foreclosed its status as natural. Disasters are disasters as such not merely because of the natural conditions that produce them, but because levees fail, governments slow-roll relief, and systems of differential governance and oppression allow and enable the traumas that environments can inflict to be experienced in radically different ways by different people. This is all to say nothing about the precise ways that climatic environments and events are not even quite natural in an era of anthropogenic climate crisis, giving lie to even the last bastion of the coherency of the concept. The hurricanes are getting worse! To take seriously the principles of environmental justice is to unequivocally commit to this premise: there is no such thing as a natural disaster.

And so, my PhD work proceeded in an iconoclastic fashion appropriate to the subject matter, not by tearing down statues and monuments, like so many did in 2020 during the protests after the killing of George Floyd and like the city of New Orleans did in limited fashion starting in 2017 when they removed a few confederate monuments,[1] but by tearing down monumentality itself. I believed—and still do—that Hurricane Katrina demonstrated that the fixation on monumentality by scholarship on public memory not only misses the functions of public memory not rendered in bronze or stone, but also is itself a symptomatic discourse of contemporary mourning. The presumption that public memory "happens" primarily at those sites is, frankly, a self-authorizing thesis that *describes* more about one's conceptualization of memory than it does directly *theorize* it. I felt that the object-fetish focus of memory studies demarcates in advance what public memory is posited to be, and I was determined to demonstrate that we can think memory otherwise, that the rhetoric of public memory also or even necessarily happens elsewhere, as it were, in unlikely and symptomatic discourses. I turned to the tools of psychoanalysis, specifically the concept of displacement, to bludgeon the point. Any reader wanting a full-throated defense of that position will find it there.[2]

This book is considerably more subtle in its approach to the question of public memory, but its thesis absolutely requires these two premises: that there is no such thing as natural disaster, and that public memory is indexed in discourses that don't claim it. In fact, the two are inextricable, for the appearance of the "natural" in relation to disaster involves a series of post-disaster rhetorical displacements that retroactively modify the concept of disaster itself.[3] I want to suggest that the rhetorical production of the fantasy of the natural disaster indexes public memory insofar as it produces the meaning of disaster itself. The conviction that Katrina was a natural disaster is an effect, a normative formalization that the global visibility of the machinations of capitalism, particularly racial capitalism in the United States, exposed—to a limited degree—as a lie. The means by which that fantasy is rhetorically constructed is what first concerns me here.

But that thesis still begs the question, which is to say, still does not prove its premise: what is the agency—and implicitly, what are the cathectic mechanisms—for this displacement? Suggesting that public memory is indexed in discourses that don't claim it raises two nonexclusive possibilities: first, it could require that scholars dealing with traditional public memory objects treat them primarily as *discourse effects*. That is, public memory is thus slightly but crucially reconceived as a process by which objects are generated that, only then, can be said to function rhetorically, epideictically, politically, and the like in the way we conceive of that discourse genre to function. The logically and chronologically prior process of generative selection—not merely through the outright public controversies over memorialization and its forms, but rather in the enunciative process of displacement itself[4]—becomes the central point of critical attention and is understood to operate in fundamentally rhetorical terms. This possibility is explored later in the book, but is preceded by the more architectonic question of how memory can be said to travel through those "places" of non-memory. The second possibility, then, involves interfacing the concerns of the public memory literature with those of the sociological and historical literature debunking the notion of natural disaster. In sum, the vast majority of the descriptive and analytic demystification of the so-called naturalness of disaster, while resolutely correct in laying out a framework for bringing the social to bear on every event which will have been named a disaster, does not often adequately address the rhetoricity of that process of naming or its possible relation—particularly given the violence of its subject matter—to public memory. At some level, it is understandable that such literature, almost all of which is produced in social scientific fields, would privilege the concepts and explanations closest to those fields. But when the pressing paradox is nominal in form (and it can absolutely be argued that the problems of sociology and disaster science and disaster economics are derivative of this problem of naming), a full theorization of the non-naturality of natural disaster requires direct confrontation with that problem of signification and thus of the rhetorical mechanisms by which the fantasy structure of the natural hides its contraband. Collective displacements of memory are not incidental, and if the trauma of disaster is experienced collectively—not in the sense that everyone experiences disasters the same, because they don't, but in the sense that they are widely disruptive discourse events that shape the trajectory of the collective unconscious through the rigorous establishment of each disaster's signification—then through some configuration it will be indexed collectively. How and where? The answer to that question cannot be generalized except in highly contextual terms, and in very loose view of the entire problematic of signifying disaster. In short: the impossibility of signifying disaster, particularly in the context of a category of disaster that appears uncontrollable and thus unimpeachable, calls for a method that traces the psychic life of public memory by reading attempts to signify it.

The crux of the methodological problem then, is fairly straightforward: if we want to know about the public memory of natural disasters throughout history, what does the rhetorician read as their text? Each of the many possible answers to this question—governmental discourses, photographic discourses, music, journalism, essays, letters, and many more texts—provide an indirect window, so to speak, into the feelings and reactions and patterns of reasoning of each moment, and this project draws on those discourses to help craft a description of these discourse formations. As far as the rhetorical work of collective memory goes, however, it is less clear how these discourses contribute actively to the public memorialization of natural disasters, which involves the epideictic rehearsal and repetition of the event channeled through objects of public memory. The simple conclusion of this line of reasoning is that a revised conceptualization of the relationship between public memory and its objects is needed, and thus also a new way of conceiving of the object domain of public memory, so that the rhetorician has something to read. This is the principal methodological intervention of this monograph.

Over the course of my broad attempt to locate memory for this study, I was led again and again to scientific modeling for the simple reason that, as a discourse, models are simulations that work across temporalities, transforming the past into the future through data consolidation and prediction. There is a certain irony at work in this configuration: predictive science posited as a mechanism by which to manage the past. As I detail throughout the chapters, these simulations take on many rhetorical forms, all of them traceable across a larger milieu of science, engineering, disaster, race, mourning, and memory. While the process of scientific development is a continual and multi-channeled one, I have partitioned several of the most widely influential forms of modeling in view of the development of disaster modeling throughout the twentieth and early-twenty-first centuries, which makes the book basically chronological. To that end, I treat the temporality of the development of these technologies, in method, as a rhetorical genealogy: a description and analysis of the major rhetorico-epistemic forms, traced over time and their (dis)continuities, and set firmly in relation to broader epistemic and rhetorical formations.

Foucauldian-inflected genealogy as a method within rhetorical studies is less a method of historical inquiry than its progenitor. Foucault designed his influential genealogical methodology in part to resist the tendency among historians to treat history, particularly the history of knowledge, as a continuous sequence ticking inevitably from one event to the next, spawning and evolving new forms of knowledge along the way. To treat history as discontinuous, on the other hand, is to recognize two things. First, that the historical sequence as such is an effect of the retroactivity of historical method—continuity emerges along lines of intelligibility and the sense-making functions within discourse domains, so signifiers will always line up later on to connect A to B, especially when the practice of writing history is called upon specifically to do so. Second, mapping this discontinuity over time can tell us something about the formations of discourse specific to each moment along that timeline. This gives the work historical specificity. In rhetorical studies, the central concern is not historical specificity *per se* but, rather, is what tracing rhetoric—its forms, its signifiers, its meanings, and its purposes—across discontinuities in time can teach us about rhetoric itself by situating those characteristics within a strong critical context. In this way, although the method takes cues from history, and the objects of study occurred some time ago, the core aim of this book is to theorize the contemporary intersections of disaster science, public memory, and environmentalism within disaster discourse by tracking its rhetorical progression throughout the twentieth century.

In order to construct the argument, I draw heavily on psychoanalytic insights to inform my reading strategy. The concept of displacement is a cornerstone of Freudian thought because it helps explain how dangerous or painful psychic material is reformulated by the unconscious so as to protect the subject from its harmful effects, even as it insistently repeats itself in the subject's memory. Freud's own thinking on these processes evolved considerably throughout his life, to say nothing of the mountainous contributions and revisions made by his many readers, and this book will not be principally concerned with theorizing Freud *per se*. But the core framework that displacement provides remains both specific enough and flexible enough to begin making sense of how forms of public memory might emerge in discourses that do not necessarily look like or claim to be concerned with public memory. With other rhetoricians who have drawn likewise from the Freudian legacy, throughout the monograph I map these psychic concepts onto rhetorical figures. Chapter 3 marries sublimation and similitude, Chapter 4 brings together different conceptualizations of transposition, and Chapter 5 conjoins Freudian formulations of anxiety, melancholy, and incorporation to theorize the very formation of memory objects themselves. In short, psychoanalysis gives me a framework and the outline of a rhetorical method for reading scientific discourses as discourses of public memory.

To summarize the overall framework: I read the broad sequence of disaster modeling science in Foucauldian genealogical fashion, and I read the discourses themselves as a Freudian rhetorician. This might imply that the book is overly theoretical in nature. On the contrary, I have quite deliberately designed its architecture to avoid this: Chapter 2 contains the bulk of theoretical work and will be where I unpack what is effectively a rhetorical theory of models and simulation, how that interfaces with and intervenes into the field of public memory studies, and what conceptual implications that conjoining has for the study of environmental communication. There, I stitch together a discussion of modeling as a scientific-yet-rhetorical discourse, the "natural" as a rhetorical effect, and the problem of public memory's objects. At the broadest level, I argue that models are memory: the rhetoric of the science of prediction, through a progression of different simulative forms, is a means of collectively working through the traumas of past disasters by projecting hypothetical ones into the future. This processing of disaster, in displaced forms, helps constitute and sustain the fantasy of the natural.

The primary argument of this chapter is that conjoining Foucauldian genealogy and Freudian displacement methodologically is a means by which to answer the following questions pertinent to the study of public memory in contemporary rhetorical studies: (1) What do we mean when we say that scientific models are a rhetorical form of public memory, and what does doing so accomplish? (2) How does one study public memory when its objects appear to be missing? (And, as consequence, what does this tell us about how these objects are produced and how we should read them henceforth?) (3) In an era of increasing natural disaster events and the ascendence of computer simulations and algorithmic prediction technologies, what conceptual tools are needed to make sense of the evolving discourses of tragedy and memory in the wake of disaster events? For the less theoretically-interested reader, this chapter can be skimmed and the subsequent case study chapters will be perfectly understandable. The experience of those chapters' claims will be richer for having engaged in the theory, of course, but the writing is meant to be accessible (even as it is sometimes very technical, because the discourse is sometimes very technical), narratively-guided, and specifically crafted in the public address style of rhetorical scholarship.

Chapters 3, 4, and 5 constitute the bulk of the textual work through rhetorical analyses of these forms and their memory. Chapter 3 encompasses the first case study: The Mississippi Basin Model. The high modernist desire to discover the secrets of nature, and thus to control it, encountered one of its limits during the early-twentieth-century attempt to tame the Mississippi. This resulted in a political debate over not only the establishment of a National Hydraulic Laboratory, but also over the prevailing scientific methodologies of disaster prevention. When one engaged in the science of disaster prevention and engineering, was it defensible and/or preferable to unlock nature's secrets through abstracted, theoretical science, or to use empirical methods to ensure that one observed the actual, "on-the ground" conditions? In resolving this tension by developing a new epistemic formulation of scientific modeling, the ensuing debate functioned also to negotiate the scientific representation of disaster itself and its position within the natural—ironically, in the attempt to eliminate them. I track physical disaster modeling through this short but consequential period, focusing on the concepts of *similitude* and *sublimation*, and arguing that the formation of this new disaster science regime also sketched the parameters of scientific representation of disaster and nature.

In Chapter 4, I turn to a scientific form that emerged as the physical modeling era began to wane: standard project storms. Starting in the 1950s and extending throughout the early-to-mid-Cold War era, a primary means for determining disaster conditions against which the U.S. Army Corps of Engineers (and subsequent federal organizations) could plan mitigation and response infrastructure was the use of what was called "standard project" methodology. The Corps used statistical analysis to craft "standard project" storms, floods, or hurricanes, which were hypothetical, statistically justifiable baseline storms that they would then transpose over an existing region, making formula adjustments based on basin shape, soil quality, average rainfall, and many other meteorological and environmental factors. In this chapter, I use a dual psychoanalytic and rhetorical formulation of *transposition* to trace these scientific discourses alongside another, simultaneous discourse domain: nuclear attack planning, which imagined hypothetical nuclear attacks by transposing pinpointed "ground zeroes" over various U.S. cities, documenting the destruction cartographically and often narratively. I argue in this chapter that transposition figured disaster modeling across discourse domains at this time, turning storms into bombs and *vice versa*, encapsulating histories of disaster into movable, deployable, and dangerous entities presented as extrinsically threatening to the nation.

As far as pure technical capacity is concerned, the cartographic and statistical era of disaster modeling was fundamentally pen-to-paper and relied on emergent computer technology only for relatively small computational tasks. As computational capacity increased significantly and then exponentially during the late twentieth century, simulation and its epistemic function transformed—radically though not fundamentally. Chapter 5 focuses directly on the first and most contentious and traumatic moment for modern, computer-simulated disaster modeling in the United States, Hurricane Katrina. The Gulf Coast region, and New Orleans in particular, has long been the site of iterative narrations of "The Big One"—a storm so brutal and devastating that it occupies a place of singularity in the national imaginary. In the decades leading up to Hurricane Katrina, disaster modeling became fully computerized. While some of their features were more or less digital translations of previous modeling technologies, one part of the method was deployed in a new fashion: repetition. Now, instead of merely using historical data to calculate disaster's parameters, modelers used that data to train algorithms that subsequently were run hundreds, thousands, or even millions of times to simulate storms. This method produced its own kind of "standard project" storm, but one that relied explicitly on the ethos of computational repetition for its legitimacy. One such storm was Hurricane Pam, a fictional hurricane designed and deployed in early 2005 for federal training purposes. The training exercises that Pam became part of predated Katrina by mere months and managed to predict almost every major meteorological dimension of its real counterpart except one: in the Pam scenario, the levees held. Absent that condition, Katrina by most accounts became The Big One. Compared to the others, and in addition to reinforcing the genealogical tracing of how models have functioned as memory, this chapter also theorizes most directly the manner in which the traumas of disaster, and in this case the anxieties of The Big One as a rhetorical formulation, are displaced onto the memory objects of those events: the Hurricane Katrina Memorial and the ubiquitous bus tours that circled the wreckage of the storm.

An undercurrent of the analysis throughout, something that should heretofore go without saying, is that any examination of the rhetoric and economics of disaster and its prevention in the United States cannot be understood but through the logics of racial capitalism. Indeed, as Robert Mizelle argues, "environmental disasters provide some of the most telling and revealing snapshots of our racial fabric across time, offering an invaluable lens into the meaning of race, citizenship, and the environmental world"[5] I recently had the pleasure of speaking with Jon West-Bey, one of the curators of the forthcoming Urban Civil Rights Museum in Harlem, NYC, a museum whose primary mission will be to explore the history of race and racism in urban contexts with a particular focus on Northern cities. Curatorial explorations of the experiences of Black Americans tend to focus, understandably, on the South, on slavery and the slave trade, and on Jim Crow and its influence on Black migration northward. The Urban Civil Rights Museum positions itself as a response to these common and well-known areas of historical concern, focusing primarily on Black urban life in the North throughout U.S. history.[6] And yet—in developing the archival and narrative scope of the project, it was insisted that Katrina be included as a resolutely racial urban event. This is of course obviously right, even if the problem of memorializing Katrina has been fraught for the almost twenty years prior to this moment, and indicates that Katrina may no longer be precisely a natural disaster in the national imaginary. With Katrina specifically, it has become axiomatic that not only was it to a large degree an unnatural disaster; it was also a racial one.

But this logic must also be extended to less remarked-upon historical disaster events: when legislators demanded that engineers develop a solution for catastrophic flooding, and when farmers and plantation owners made demands of legislators concerning what should be the federal obligation to maintain such infrastructure, and when this obligation was rooted primarily in the national reliance on the preservation of agricultural lands, the great (mostly) unspoken engine guiding this sequence was systemic mass exploitation of labor, cut as it necessarily was across racial lines in the United States. Later, when the racial geography of Ground Zero became evident in the atomic age, when Hurricane Camille tied disaster relief to public school funding, when the memorialization of Katrina reenacted and voyeurized Black misery in New Orleans, legislators and scientists continued framing these events as natural disasters in need of technological mitigation. To do so in the face of such a clear pattern complicates the thesis, explored in the next chapter, that modernity transformed disaster into an administrative problem of governance—unless, of course, one accedes that race is a constitutive part of modern capitalism, and that the self-evidence of "nature" in disaster is coextensive with the self-evidence of racial difference. As Barbara Fields writes of the Founders in her highly influential essay "Slavery, Race, and Ideology in the United States of America," "Those holding liberty to be inalienable, and holding Afro-Americans as slaves, were bound to end by holding race to be a self-evident truth."[7] And, as Destin Jenkins & Justin Leroy have made clear, adopting the lens of racial capitalism involves a refusal to see race as "anterior to deracinated capitalism" but, rather, demands that any analysis of capitalism must also involve an analysis of the way that capitalism constitutes racial difference in its historical particularity.[8] Insofar as natural disaster mitigation is animated by a minimization of negative impact on systems of economic production, it is animated by the production of race at the heart of that system. Thus, while the analyses herein are not organized explicitly by the question of race, race does and must animate the analysis throughout.

While the analyses are more centrally organized around modeling technologies, limitations on scope are inevitable. There are many, many different forms and practices that one could include in both the categories of models and simulations. The object domain of my attention is not meant to be a comprehensive survey or charting of all the various rhetorical configurations of disaster modeling. The closer to the contemporary moment we find ourselves, in fact, the proliferation of modeling and simulation—computational proliferation, to be sure, but also in popular culture, organizational exercises, and news media—makes distinguishing clearly between all the forms and arguing their functions nearly impossible to do with any confidence. Rather, even as the technical details of these objects dominate the pages herein, the underlying concern remains focused on theorizing public memory in the context of natural disasters and the indeterminacy of disaster's representation through rhetoric. Concentrating attention on the objects, methods, and practices that help us answer those questions was the primary objective in constructing the architecture of this book.

As outlined in the preceding pages, my gambit in this book is that the science of predictive disaster modeling indexes public memory and the relationship between disaster and collective trauma because, in specifying and replicating the conditions of believability, the patterns of acceptability, and the acceptable levels of vulnerability, various technologies of modeling issue protocols for disaster regimes that prefigure their effects. These prefigurations have two primary features. First, they prefigure the intelligibility of disasters to come—even, or especially, when they are mostly accurately deployed. Second, they themselves are positioned as a formation of memory-work. Their emergence, one by one, comes regularly in the wake of actually occurring disasters. As representations, approximations, *simulations*, their rulesets and features and scales and properties and assumptions require an authorizing appearance of "realism" that folds back, chiasmatically, to verify their integrity. That is to say, predictive models achieve their veracity through strategic distortions that resonate with empirically observed patterns and sequences. When scholars argue that public memory is very much a concern for the present rather than the past, even in discourses about our future, this temporal whiplash is necessarily implied.

NOTES

1. Jeremy R. Grossman, "Take 'Em Down: Rhetorical Temporality and Critical Regionalism in the Struggle to Remove Confederate Monuments in New Orleans," in *Reconstructing the South: Critical Regionalism and Southern Rhetoric*, eds. Christina L. Moss and Brandon Inabinet (Jackson: University Press of Mississippi, 2021), 126–43.

2. Jeremy R. Grossman, "Dreaming of Disaster: Displacements of Public Memory and Hurricane Katrina" (Dissertation, University of Georgia, 2018).

3. To avoid the stylistic burden of the proliferation of scare quotes, I have chosen to omit them throughout this book. May the reader know, however, that the conceptualization of the "natural" henceforth involves the consolidation of the natural as fantasy structure rather than as an ontic reality.

4. Here I use "enunciative" in the Lacanian sense of the term, *énonciation*: speech read not at the level of its content or meaning, but rather at the level of its having been uttered in its performative specificity.

5. Richard M. Mizelle, Jr., *Backwater Blues: The Mississippi Flood of 1927 in the African American Imagination* (Minneapolis: University of Minnesota Press), 24.

6. "National Urban League's Innovative Urban Civil Rights Museum Will Bring Little-Examined Corners of History to Light," National Urban League, 12 July 2023,

https://nul.org/news/national-urban-leagues-innovative-urban-civil-rights-museum-will-bring-little-examined-corners.

7. Barbara Jeanne Fields, "Slavery, Race, and Ideology in the United States of America," *New Left Review* 181, no. 1 (1990): 101.

8. Destin Jenkins and Justin Leroy, "Introduction: The Old History of Capitalism," in *Histories of Racial Capitalism*, eds. Destin Jenkins and Justin Leroy (New York: Columbia University Press, 2021), 10.

Chapter 1

On Models and Memory

As sketched in the introductory chapter, there is an explicitly practical question animating the theoretical life of this book: What is the public memory of natural disasters? Because there are often no memory objects, no practices or places of commemoration for what are quite often horrific events, one must read elsewise and elsewhere. The sequence of the present chapter moves through three theoretical nodes that sketch out a framework for answering this question: public memory and its fixation on memory places and memory objects; modeling as a conceptual technology of prediction; and the fantasy of the "natural" disaster that animates conversations around climatological catastrophe. In moving through each node, one by one, I sketch out a rhetorical genealogical method for reading public memory as well as a conceptual justification for bringing these diverse concepts and fields of study into conversation. Developing a satisfying answer to the problem of public memory's absence requires a sensitive framework, one that can assimilate the insights of public memory studies while avoiding the tendency as critics to find exactly what we're looking for, exactly where we expect it to be.

PUBLIC MEMORY AND ITS OBJECTS

The overwhelming majority of objects of analysis within public memory scholarship have at least one thing in common: they are objects of public memory that are self-referentially identified as such. That is, they are designed specifically to appeal to memory. Pierre Nora's landmark volumes on French national memory, originally published in French under the title *Les Lieux de Mémoire* and translated into English as *Realms of Memory*, began as the attempt to tell the history of France through what in this book I will refer to as memory objects: memorials, commemorative rituals, slogans, museums, and others. As he notes in the preface to the first volume, "[o]nly certain works of history are *lieux de mémoire*, namely, those that reshape memory in some fundamental way or that epitomize a revision of memory for pedagogical purposes."[1] This is the rationale for the object domain of the first two volumes. By the third and final volume, however, the term *lieux* came to encompass a breadth of objects of analysis that reflected the ambiguity of the word in French and included not only manifest memory objects, but also "latent or hidden aspects of national memory."[2] The domain expanded to include philosophers, historical figures, and even the French language itself, a recognition that the mechanisms of collective memory may also inhere at the level of structure rather than solely via the authorization of memory's objects. Here, too, however, each of the analyses tend to highlight encomia, commemoration, physical sites of memory, and other rhetorical forms that we would recognize straightaway as self-referentially memory objects. The volumes together, in the attempt to tell France's history through its memory objects, operate however loosely on the assumption that memory objects are the principle *lieux de mémoire*.

Central to his distinction between memory and history, then, is the observation that memory is a thing of the present, rather than the past: "Memory is always a phenomenon of the present, a bond tying us to the eternal present; history is a representation of the past."[3] While the strict epistemological division between history and memory is one that scholars have since challenged,[4] literature on public memory typically retains that memory objects function within the present and are invented "in order to assign a meaning to the past that accorded with their [creators'] contemporary concerns."[5] The negotiation between past and present that this outlook implies, according to Kendall Phillips, has both challenged the authority "of a fixed, singular history" and "highlights the extent to which these constituted and constituting memories are open to contest, revision, and rejection."[6] For these scholars, the malleability of memory positions it as a privileged object domain the study of which is paramount to describing the rhetoric and politics, not of the past, but specifically of the present's past.

This malleability, however, has engendered the anxiety of "misremembering" at least as far back as Aristotle and Plato. In his short treatise, "On Memory and Reminiscence," Aristotle lays out what is basically a Platonic conception of the difference between memory (*mneme*) and recollection (*anamnesis*). For him, memory "involves a presentation: hence we may conclude that it belongs to the faculty of intelligence only incidentally, while directly and essentially it belongs to the primary faculty of sense-perception."[7] Insofar as memory relies on presentation and image, not only is it an "emotional state" for Aristotle, as Phillips has observed,[8] but it is also the case that one can misremember because of this fact. In illustrating this principle, Aristotle references people like Antipheron of Oreus who, "suffering from mental derangement . . . were accustomed to speak of their mere phantasms as facts of their past experience, and as if remembering them. This takes place whenever one contemplates what is not a likeness as if it were a likeness."[9] Whereas recollection is a systematic, disciplined ordering of memory grounded in rigorous intellectual activity, memory is an "affection" that holds no inherent relation to truth—indeed, for him, this is precisely why animals can have memory but not recollection.[10] For Phillips, the separation of memory and recollection signals an animating anxiety within contemporary public memory studies: "the failure of our memory through the process of misremembering or misrecognition."[11] He notes, "Plato's concern for misremembering is driven not so much by a pure concern for the epistemological or phenomenological foundations of memory as it is by a broader concern for our capacity for false beliefs."[12] And while Aristotle turns to an example of a person with "mental derangement" to describe false memory, the structure of his argument holds that this happens to everyone to some degree, not merely the mentally deranged. This is the fallibility of the relationship between sense-perception and recollection.

Yet, for rhetorical studies, misremembering, misrecognition, and forgetting are often taken as that which is most instructive about memory and has enabled scholars to describe the ways in which memory objects and rituals routinely function in ways not overtly expressed in the content or form of the object. Rhetoric's attention to form, trope, affect, argument, and identification in the production of social and national collective identities casts public memory objects as the means by which the historical significance of a particular event is constituted in the present. This approach sees public practices of remembering and forgetting not simply as reflections or representations of present-day ideologies, desires, affiliations, anxieties, and/or investments, but also as performances that reproduce them. For rhetoricians, objects of public memory have signaled more than simply the management of the past through the lens of the present or the distortion and appropriation of history for the present; they are constant reaffirmations of the present through the repeated constitution of the present's past.

The instability in the relationship of public memory objects and the past has prompted the theorization of not only the specific truths that such objects espouse, but also of the emergence of particular objects of memory themselves. Why do certain memory objects appear and not others? In perhaps the most widely cited and comprehensive synthesis and critique of public memory studies within rhetorical studies, Blair, Dickinson, and Ott offer one answer, noting that "groups talk about some events of their histories more than others, glamorize some individuals more than others, and present some actions but not others as 'instructive' for the future."[13] Their answer to the above question—Why this memory object and not another?—involves a methodological turn to Michel Foucault's early work on what he calls "rarity."[14] Out of the plethora the possible things that can be said within any historical context, certain things end up being said and others do not. It is therefore the task of analysis to determine the "law of that poverty," the laws within any discursive field that rarify the discourses found in it.[15] In so doing, the memory scholar's task is in part to outline the rationality within which memory objects emerge, the intelligibility of the forms that memory objects take, the events they ask us to remember, the ways in which we are to remember them, and the forms of belonging that accompany these narrative appeals. In short, it is to combat the risks of both anachronism and Platonic/Aristotelian idealism head on by describing the discursive contexts within which memory objects emerge and within which their epideictic appeals become meaningful.

Part and parcel of the Foucauldian answer to how memory objects emerge is the precept of immanence, that they emerge in society from within systems of power and knowledge and not due to some transcendent structure. As such, scholarly inquiry into that which does not appear within these systems of relation makes the grave mistake of positing a transcendent structural plane. Foucault argues that the "why this and not something else" question must not be mistaken with the presumption of latent, unrealized discourse: "we are not linking these 'exclusions' to a repression; we do not presuppose that beneath manifest statements something remains hidden and subjacent. . . . The enunciative domain is identical with its own surface."[16] This repudiation of Freud and structuralism writ large rejects the idea of an unsaid underbelly of discourse, an architectonic of repressed truths that would act as a constitutive condition outside the field of discourse. Joan Copjec summarizes:

> He correctly and strongly believes that the principle of a regime's institution cannot be conceived as a *meta*principle, that is, as a logical observation that is simply added to all the other observations one may make about a particular regime in order to organize, embrace, or comprehend them. The principle of construction or staging cannot occupy a different, a superior, position with respect to the regime it stages.[17]

Thus, for public memory, it would seem a logical inconsistency to pursue a discourse of public memory outside of the nominal objects of public memory, for so doing would seem to imply the existence of just such a metaprinciple. What internally coherent logic would allow a discourse to be counted as a *lieux de mémoire* except for its nominal designation as such?

Copjec turns Foucault's framework back on itself, however, to suggest that, not only is he unable to complete the task of defining "the *constitution* of domains of objects and knowledges" without some sort of "a principle or a subject that 'transcends' the regime of power he analyzes"[18]; more importantly, it unwittingly installs the acknowledgement that "the whole of society will never reveal itself in an analytical moment . . . no diagram will ever be able to display it fully."[19] In other words, because Foucault will not admit into his analytical framework anything outside a given regime of power, he gives up in advance the possibility of describing the conditions of emergence for said regime. Her Lacanian corrective to this is to say that, in fact, "what we do when we recognize the impossibility of metalanguage is to split society between its appearance—the positive relations and facts we observe in it—and its being, that is to say, its generative principle, which cannot appear among these relations."[20] The inevitability that language can never speak of itself fully ensures this split and provides the claim, not that society can never transcend its own discourse but, rather, "that society *never stops realizing* itself, that it *continues* to be formed over time."[21]

The same can be said of memory: because of the impossibility of pure recollection, which Aristotle so vehemently sought to achieve, memory never stops realizing itself, never stops being formed over time. Misremembering and misrecognition, then, as Aristotle knew when he located the mental disturbance always threatening recollection, are fundamental not only to the content of memory but also the very objects it "chooses." Thus, rarity—why this memory object and not another—is not what is at stake here. Rather, what is most central to an analysis of public memory's displacements is that generative principle which has the authority to institute an ever-changing domain of objects that "count." By extension, this leaves open the possibility that the truths that public memory provides, the meaning and significance of past events, may well also generate (in) *lieux de* non *mémoire*, realms outside of the intelligible discourses of public memory.

To carry out a reading based on this premise, it will first be necessary to outline the distinction between the subject of dreams and the unconscious as Freud articulates them. In *The Interpretation of Dreams*, Freud begins the all-important chapter on dream-work with an observation:

> Every attempt that has hitherto been made to solve the problem of dreams has dealt directly with their manifest content as it is presented in our memory. All such attempts have endeavored to arrive at an interpretation of dreams from their *manifest* content or (if no interpretation was attempted) to form a judgment as to their nature on the basis of that same content.[22]

Dream interpretation had historically centered on the actual content of dreams, attempting to interpret the meaning and function of dreams based on the images, narratives, people, etc., within them. In opposition to this approach, he suggests that the key to the interpretation of dreams lies in the interpretation of the dream-work involved in *generating* the manifest dream-content, or what actually makes it into a dream. This dream-work has two fundamental forms: displacement and condensation. These are names for the dream-work, or the psychic activity of transposition, that determines the content of dreams on the basis of the structure of the unconscious. Displacement does so by "strip[ping] the elements which have a high psychical value of their intensity, and . . . creat[ing] from elements of low psychical value new values."[23] Condensation involves a "mass of dream-thoughts being submitted to a sort of manipulative process in which those elements which have the most numerous and strongest supports acquire the right of entry into the dream-content."[24] That is, those dream-thoughts with the greatest psychic intensity are condensed repeatedly and multiply into associated, but screened, dream-content (this is his definition of "over-determination"[25]). In fact, the modern dismissal of unconscious processes in favor of positive content, Julia Kristeva argues, is motivated by a (deliberate?) ignorance of the scope of Freudian philosophy: "These destroyers of Freud ignore that, far from reducing psychic life to the organs that perform the sexual act, psychoanalysis understands if and how excitation, pain or pleasure, is integrated in the complex architecture of sensations, speech [*paroles*], thoughts, schemas [*projets*]."[26]

As Christian Lundberg notes, Lacan takes up this division by translating displacement and condensation into the rhetorical tropes of metonymy and metaphor, respectively, thus securing for rhetorical studies a warrant for theorizing the rhetoricity of the unconscious: "[t]he logic that inheres in dream work is the same logic that underwrites the function of speech generally."[27] It is this shared logic that prompts Lacan to endorse a theory of language grounded in the subject of enunciation, or the subject of the unconscious.[28] The latter may be thought of as the commonplace first person singular, the "I" grammatically positioned as the subject in a sentence or, in the case of public memory, as the first person plural "we" implied in consubstantiality. This is the subject of the ego or consciousness, the positive "we" produced, for instance, in collective memory objects that constitute publics or peoples. Enunciation, on the other hand, concerns the unconscious, which I take to be the rhetorical site of the (re)production of memory.

Distinguishing between the subject of the statement and the subject of enunciation is not only a theoretical shift, but also a methodological one. Lacan sums up the difference: "We can try here, in a concern for method, to depart from the strictly linguistic definition of *I* as signifier, where it is nothing but the shifter or indicative that in the subject of the statement designates the subject insofar as it is currently speaking. . . . That is to say, it designates the enunciating subject, but does not signify it."[29] Whereas the subject of the statement is what we would typically recognize as one's conscious identity, the subject of enunciation is that which statements mark out; it is the subject of the unconscious. To follow Lacan is therefore to presume that the "I" of the subject "is born somewhere other than the place where the discourse is enunciated."[30] A fuller theoretical account of this will unfold in the chapters to follow, wherein the same premise is assumed in analyzing public memory: the presumption that the contents of public memory, i.e., its objects, are born somewhere other than in self-reference, and that the psychic functions of public memory are sometimes carried out in discourses outside of the object domain of memory culture.

Psychoanalytic scholarship in public memory studies has to a degree already taken up Barbara Biesecker's charge to reroute the analysis of rhetoric through a Lacanian psychoanalytic apparatus. Her own work regarding WWII memorials, for instance, argues that the reemergence of WWII as particularly resonant object of memory coincides (not accidentally) with the transition from one era into another, from one relationship of the subject to itself and another. Likewise, Claire Sisco King's allegorical reading of *Poseidon*, to which I return more fully below, posits the film as a form of screen memory, or memory-work that "screens off" certain traumatic histories, for the emasculating trauma of 9/11.[31] Finally, Joshua Gunn's engagement with the Huey Long monument in Baton Rouge considers its uncanniness in the context of Long's demagoguery, which he argues is situated in the dialectic between obsessional neurosis and hysteria.[32] These essays and others take as their starting point enunciation itself, the specific contexts in which public memory of some event must be worked through for reasons seemingly independent of the form and content of the objects themselves. In accordance with this approach, it is only with a primary focus on enunciative displacements—that is, both the way that objects of public memory are themselves displacements and that memory-work functions outside of that domain—that attention to the content of these objects begins to elucidate the collective subject that they are said to produce.

Yet, the object domain of rhetoricians theorizing public memory through a psychoanalytic framework tends to remain basically consistent with the field of public memory studies as a whole. Even among those scholars who read memory as an enunciative practice, the assumption that traditional *lieux de mémoire* constitute the arena wherein public memory is *happening* insists: museums, monuments, memorials, commemorative events, historical sites, and so on. As such, public memory's object domain has in many ways determined the limits of the theorization of its rhetorical function because it is presumed from the start that these are the primary—sometimes even the exclusive—sites for the production of memory.[33]

In suggesting that rhetorical scholars theorizing public memory as a whole have tended to limit their object domain to the manifest content of public memory, and in reappropriating Freud's work on dreams as grounds for this claim, the architecture of this book proceeds entirely under the presumption that rhetorical memory-work, like dream-work, consists of rhetorical forms that can be traced and analyzed through these displaced objects. Readers who cannot abide this presumption will, unfortunately, find in this book a very limited quantity of pages devoted to the question of public memory as it has been traditionally formulated. Readers who are willing to adopt this framework, even provisionally, will find memory-work even in the most technical and scientific portions of this book, for this inquiry moves well outside specific attention to remembering and forgetting and towards their dialectical relationship, which I take to be constitutive of memory as such.

Reading this constitutive dynamic through a psychoanalytic frame allows the critic to trace the specific, enunciative organization of this threshold. How so, if Bradford Vivian is correct in asserting that "[o]ne may cogently describe the *raison d'être* of psychoanalysis—to uncover and explain the secrets of the unconscious—as a project against forgetting. . . . Forgetting is psychologically unhealthy: it leads to repression, which leads in turn to debilitating neuroses or psychoses"?[34] Vivian's project presents a formidable challenge to the remembering-centered tendency of public memory studies and argues that calls to forget also serve public, political functions in the constitution of memory, and therefore rightly places Freud's work within a larger modernist project invested in remembering.[35] Psychoanalysis as both a method of treatment and a critical heuristic, however, incorporates forgetting conceptually as a necessary function of the psyche. Indeed, if we understand forgetting to be descriptive of all the ways of not remembering, then Freud can be said to have multiplied the field of forgetting into all the defenses, not merely to goad the patient towards remembrance, but also to describe the manner in which subjects manage the traumas of their histories. Grounding an inquiry at the threshold of this dialectic is therefore not meant as a project against forgetting *tout court* as much as it is a criticism of the rhetorical function of certain configurations of forgetting and remembering. Memory-work is descriptive of nothing but this relationship.

Rhetorical scholars have, at times, approached the question of memory at the site of this dialectic. For example, Claire Sisco King's analysis of *Poseidon* as an allegorical repetition of the trauma of 9/11 in many ways embodies a reading of memory-work at the level of enunciation. She contends that, according to Freud, past trauma must be repeated again and again in the hopes of mastery of said trauma, and such repetition is screened off by way of allegory:

> [A]s much as the urge to replicate the past might expose wounds felt by the national collective, these acts of repetition may also indicate attempts to repair or reverse such damage by rewriting traumatic history. And more, the compulsion to replay and revise traumatic events can be animated allegorically in

> texts that otherwise fail to address the traumatizing violent scene in direct or explicit terms.[36]

In reading the 2006 film *Poseidon*, she demonstrates that the film repeats, or renarrativizes, the traumas of 9/11 in order to resolve post-9/11 trauma that threatened masculinist figurations of national identity in the United States. The allegorical framing aligns the film sufficiently with the events of 9/11 that it functions as a repetition, but revises the narrative in a number of ways, promising "protection from trauma's painful and intrusive return."[37] The revisions of the film, she argues, allow it to appear to say nothing, and therein lies the displacement: "For Freud, screen memories are inconsequential, or 'trivial,' memories that displace 'objectionable' ones in acts of forgetful (mis)remembering meant to offer a defense against, or avoidance of, a traumatic past."[38] For her, it is the very displacement of the traumatic scene (9/11) that allows these memories to be repeated allegorically, for they flare up as something else (not 9/11) whose narrative logic is mostly the same, but whose content has been screened as if in a dream.[39]

To follow the temporal imperative of this sort of analysis is to say no more than what scholars of public memory have said before: public memory in its manifest content is always a matter of the present, as is scholarly analysis of it. Marking the boundaries of public memory scholarship on the basis of these objects has produced a vast compendium of insights about how they function, their capacity for mediation and making meaning, for negotiating truths, and for fulfilling (or not) the promises that epideictic speech makes to various publics. At the same time, I would like to challenge the implicit laws governing that object domain by way of an encouragement to remember differently the translation of Pierre Nora's landmark publication of *Les Lieux de Mémoire*. In his preface to the English translation, *Realms of Memory*, Nora recalls not only the trouble one has with adequately translating the French noun *lieu* into English, but also the ways in which the French term came to take on a more amorphous definitional quality throughout the course of even his own project's development:

> The narrow concept had emphasized the *site*: the goal was to exhume significant sites, to identify the most obvious and crucial centers of national memory, and then to reveal the existence of invisible bonds tying them all together. . . . The broader conception required by the planning of [the final volume] *Les France* placed the accent instead on *memory*, on the discovery and exploration of latent or hidden aspects of national memory and its whole spectrum of sources, regardless of their nature.[40]

In short, what began as a project of producing a history of French nationalism by examining a wide range of memory "sites"—memorials, "museums, commemorations, and mottoes"[41]—upon broadening the conception of "site" and putting emphasis instead on "memory" ultimately became a project of retelling France's history through its most seemingly unrelated and unremarked *lieux de mémoire*. His anxiety over this method is, of course, that some readers would be appalled, that to set an "almost caricatural symbol such as the Gallic cock alongside the palace of Versailles or the battle of Verdun . . . is to blur the distinctions between the greatest and most brilliant accomplishments of French history and tradition and the humblest instruments for fabricating that history and that tradition," and that "to do so runs the risk of appearing to diminish those accomplishments."[42] To this he responds: That is precisely the point, a French history without French nationalism. The running assumption that the monuments of our memory should dictate not only object domain but also method reflects a certain adherence—even a critical one—to the *a priori* constitution of that domain. The widening and rethinking of that domain on the basis of this insight is a central pillar of the current project, as it has needed to be for other scholars who, for instance, attempt to document the public memory of marginalized and oppressed peoples whose histories may not have been preserved and for whom traditional memory objects are not always built.

It will be my wager throughout that Freud and his readers provide a necessary part of the framework for reading such a broad and seemingly disconnected object domain of memory. Is it not the case that the interpretation of dreams is to a significant degree the attempt to discern a unifying symbolic framework out of an often ridiculous set of manifest dream content? Freud argues in The Interpretation of Dreams that absurdity as such only presents itself in dreams where the dream-work attempts to represent contradiction.[43] In all other cases, the ridiculousness of dreams, the setting alongside of the Gallic cock and the palace of Versailles, is not questioned by the dreamer, for the Freudian bargain holds that some psychic process intelligibly links them in its logic (and later, with Lacan, in its language). The task of scholars of public memory who take up this Freudian charge, then, is to do the same. While Nora's concern with object domain is a shared one in this task, the method I begin to suggest differs fundamentally. The aim of the analyst/critic—unlike Nora's historiographer—is to interpret the logic of the memory-work on the basis of its displacements, not to figure out what the cock and the palace have to do with one another.

Given my specific problem of the missing objects of memory, and in view of the basic framework of displacement that I've spent the past few pages outlining and defending, a front-and-center definition of public memory is probably in order—particularly if, as scholars of public memory have successfully argued for decades, public memory should be treated with distinction as a special kind of discourse. For this book, public memory shall be defined as an enunciative discourse, working primarily through repetition and displacement, whose proper function is to ensure the rhetorical continuity of past, present, and future; manage collective psychic responses to objects of consequence; and consolidate the epistemological and rhetorical means of remembering those objects. As is the case with every project, this definition privileges the dynamics at work in my own object domain, and so the "objects of consequence" in question are natural disasters. However, hewing rather closer to my object domain here also serves as a tempering force against a critique that might find a broader definition of enunciative public memory capacious enough to obliterate the concept itself, or at least its distinction from a general theory of the history of discourse, *des choses et les mots*.[44] I have chosen instead to uphold the presumption that there may be some distinction to the discourse of public memory after all, particularly in its functions as a post-traumatic rhetorical technology.

The more-or-less Freudian rhetorical reading strategy with which I engage texts and their metonymies is nevertheless routed through what is essentially a Foucauldian approach to history: The emergences and shifts in discourse formations, the epistemological arrangements in the formation of truth-discourse, and a keen interest in the technologies of knowledge production and their relation to institutional power. And, ultimately, the biopolitical question that insists as disasters became principally problems of the institutional management of life and death. I map these discourses, technologies, and discourse formations from the beginning of the twentieth century through the modern era (and, in the conclusion, speculatively, into the future) via a rhetorical genealogy of disaster modeling rationalities and the simulations they produce. While many different sorts of models and simulations are examined, some overview of the theoretical position of models is required to situate their epistemological and rhetorical status.

MODELS AND SIMULATIONS

First, a word about terminology: "model" and "simulation" are sometimes used interchangeably, and while the advent of highly advanced simulation technologies makes a clear distinction challenging on epistemological grounds (as we shall see), the two will be operationalized here very basically according to Aimee Roundtree's understanding:

> Although many scholars call simulations "models," technically speaking, simulations go a step further than mere models [because] simulations capture a

> model's behavior. . . . While scientists use numerical models to help establish parameters of a simulation, the models in and of themselves represent static features. Simulations, on the other hand, put models through numerous conditions, variables, and states.[45]

Roundtree's distinction is a fair and pragmatic one. Models are designed as a ruleset consisting of an internal logic and a set of parameters (even dynamic ones), and through that ruleset scientists can run any number of tests or inputs, which in turn produce different results. Those independently conducted sets of inputs are different simulations of and within the same model, all responding not only to the controlled change of inputs, but directly and more consistently to the ruleset that the model lays out in advance. Indeed, even in Jean Baudrillard's broad, almost all-encompassing conceptualization of the relationship, "[s]imulation is characterized by a *precession of the model*, of all the models based on the merest fact—the models come first, their circulation, orbital like that of the bomb, constitutes the genuine magnetic field of the event."[46] And yet, the epistemological function of the model contains the precession implicitly: in scientific discourse, the model is not an artistic conceit, it is meant to be used.

And for the truth function of such discourse to be authorized, the temporal relationship between the model, the simulation, and so-called reality must be established. For Baudrillard, simulation functions as a rough synonym for representation, summed up in the well-known Jorge Luis Borges fable "in which the cartographers of the Empire draw up a map so detailed that it ends up covering the territory exactly."[47] Representational signification, metaphorized in the fable by the comic outcome of the map achieving a one-to-one correspondence with its referent, in postmodernism had given way to a new function of simulation whose primary task is to mask the absence of the referent, which deploys simulation within its models in order to effect a disavowal of the fact that the models have in fact no basis in the real. In such instances, the question of whether a particular simulation adheres to the forms and patterns of reality is subordinated—through a chiasmatic reversal—to an assessment of whether reality remains sufficiently authentic to the model. This is *simulacra*, and most scholarly attention on this landmark work has grappled with its implications on the concept of hyperreality. But Baudrillard's anxiety over the emergence of the hyperreal and the loss of the sign, the disintegration of a metaphysics of language, produces a conceptual framework within which the role of simulation is only ever fetishistic: the obviation of the loss of the sign's relation to reality by a chiasmatic inversion of the relation between simulation and reality. Simulation can only ever be a symptomatic copy and, within the domain of the hyperreal, will always twist back around to distort reality.

But what Baudrillard's chiasmatic framework does allow for is a thickening of Roundtree's notion of the temporality of the relationship between model and simulation. In completely practical terms, the capacity, within both theoretical and empirical scientific activity, to create models and simulations that function basically in a way that mimics reality is necessary for confirming the epistemic conceits of modeling: this capacity is the primary condition for positing that modeling can produce truth. This does not suffice as a philosophical rationale, for a variety of reasons related to the assumed epistemic status of both truth and simulation (not to mention reality), but as an argumentative and practical one, it allows practitioners like engineers and urban planners to act with some degree of confidence. And in this register, models are logically and chronologically prior to simulations, internal to the process or method of the production of knowledge. As examples in this book demonstrate, a meteorological researcher must formalize the parameters of the model before they are able to run thousands or millions of iterative simulations of a flood or a hurricane. And they must specify the climatic variables and cartography of storm formation, as well as their relationship to the internal physical architecture of actual storms, before they can simulate a "worst case" storm for any given location. These processes are placed both logically and chronologically prior to the act of simulation.

Yet, the temporality of both their veridiction and their conceptual formalization remains an entirely different matter and impinges directly on their rhetorical and epistemic functions. In short, the conceptual process that authorizes models to produce truth is not necessarily linear. More than a decade prior to Baudrillard's polemical denunciation of the contemporary status of the sign, Alain Badiou began lecturing at l'École Normale Supérieure on what he called "the concept of model."[48] The treatise that resulted locates in the titular concept an epistemological problematic that deals directly with the production of knowledge and is grounded in the field of mathematics. To accomplish this task, Badiou finds it necessary to address a prominent conceptualization of the relationship between the model (which, in that it consists of the theoretical rules of relation between formal elements, he calls the "syntactic") and the empirical (which, in that it consists of the observable facts that the model attempts to index, he calls the "semantic").[49] This notion he elaborates through Claude Lévi-Strauss's structural anthropological work, which holds that the model is the technical artifice through which one deduces scientific truths about reality. Moreover, its mode of verification, that is, the means by which one proves that the model produces accurate truths, is principally the apparent resemblance those truths have to this so-called reality. This creates a circularity that Badiou will not abide. As artifice, the model attempts to *explain* the empirical, observable facts in the world, but the way in which one verifies the model is in its *resemblance* to those very same facts. Badiou laments, "[i]t is plain to see how external analogy and simulation are called upon here to reduce the initial gap between the inert opacity of the facts and the constructive activity of models."[50] That is, within such circularity, simulation and analogization rush in to authorize the relationship. Those processes do more than simply stitch up this contradiction: insofar as simulation and analogization rhetorically uphold the legitimacy of "the duality of reality and model,"[51] Badiou notes, they function as reproductions of another duality, that of "fact" and the "law," both of which are presumed to operate independent of the activities of the scientific production of knowledge. The bifurcation of something like "natural laws" and "facts according to those laws" shields the activity of science from its own constitutive role: it would follow that scientific practice merely attempts to theorize the fundamentals of these natural laws by getting the facts to line up, as it were. As the numerous case studies throughout the present book will demonstrate, the epistemic function of this reproduction remains not only the conceptual maintenance of this duality, but also the presumed objectivity and neutrality of both halves of it.

The impasse one finds here in the circularity of modeling epistemology is a crucial point of critique because it reframes inquiry away from the most direct post-disaster postmortems of such models—whether and how a given model ended up being accurate or not, astonishment at the breathtaking accuracy of hypothetical models to predict actually-occurring disasters, discussions about why these incredible models are or were not taken more seriously by some combination of institutions, and/or for whose vindication or benefit these models are produced when they are routinely ignored by said institutions. These are necessary questions, both for the intrascientific refinement of models and the interfaces that turn modeling into politics, ethics, philosophy, and the like. Moreover, it is unequivocally true that scientific modeling practices have assisted in the challenge of anticipating and mitigating the climatological effects of natural disasters, and the recognition of both this and the most general belief in the imperfectability of science is certainly enough for many to justify the history and technics of disaster modeling *tout court*. Disaster science and engineering has saved countless lives and continues to do so. But, as Sandra Harding posited so elegantly more than thirty years ago, an apologia for science that seeks to take credit for the good and disavow responsibility for any of the bad may constitute and maintain a particular scientific ethos, but it does nothing to prevent a strong, responsible, and illuminating critique of that science.[52] Whether the sciences accept the premise or not, humanistic analysis understands science to be a vast surface of discourses, epistemic formalizations, and reflections of cultural and ideological values; that science and its defenders often fancy the work epistemically incorruptible as long as its methods are sound does not make it so. Nor does it negate or deny the possible effectiveness of science's outcomes—the critique is not concerned with that question one way or another. Critical scholars studying STEM fields are almost always pursuing different questions than those of the fields under study.

In my case, the epistemic circularity of the model is the point of entry for thinking models differently: how can their rhetorical function be characterized? And logically prior to this question, what is the enunciative function of modeling itself, what does it produce and what does it displace, in psychoanalytic terms? Famously, Jacques Lacan insists at the end of his "Seminar on 'The Purloined Letter'" that "the letter always arrives at its destination,"[53] which is to say, all too loosely, that the apparently predestined—or, at least, anticipated—clicking together of a series of events into an intelligible moment of signification is something done by the subject in a retroactive mode. This confirmation is an insistent feature of intersubjectivity. It is also the first level of critique of the model: the closed system of inputs, parameters, and outputs—however dynamic they are formalized to be—spits out a set of signifiers that cannot but have been spit out. In explaining this function within the Lacanian imaginary, Slavoj Žižek provides an example:

> When physicists endeavor to reconstruct the process culminating in the appearance of intelligent living beings on Earth, they either presuppose that universe was created in order to render possible the formation of intelligent beings (the "strong," overtly teleological anthropocentric principle) or accept a "circular" methodological rule requiring us to always posit such hypotheses about the primeval state of universe as to enable us to deduce its further development toward the conditions for the emergence of life (the "weak" version).[54]

That the methodology is circular in this instance may be philosophically weak, that much is true, but it also means that certain ideological and epistemic predispositions are easily baked into the process—in fact, will almost certainly be baked into the process, for the precession of a model into a simulation involves directly catalyzing these predispositions towards a particular scientific end, the answer to one question asked instead of another question.

Scholars in the broad interdisciplinary field of science and technology studies (as well as, increasingly, journalists publishing in popular outlets), for instance, have long labored to demonstrate the irrefutably ideological character of these presumably neutral and objective technologies, and rhetorical studies scholars have labored more specifically to demonstrate that rhetorical processes are the architecture of this ideological expression.[55] In other words, against the popular perception that scientization and quantification are mechanisms of objectivization, that the data don't lie, scholars in the humanities and social sciences have argued the numerous ways that models and simulation—both as a technology and an epistemic method—cannot be seen apart from the world from within which they emerge. Roundtree, for instance, notes the ways in which models and simulations can be "biased" or distorted within in their own internal logic by, for instance, "ad hoc reasoning: making assumptions, substitutions, and following hunches."[56] In fact, she notes a moment later, they *must* be: "Simulation scientists must falsify future hunches or hypotheses about what will happen, rather than what phenomenon is happening now or did happen in the past. In this way, simulation scientists must judge not the present truth but the potential truth of simulated data or findings."[57] The effect-structure of this predispositionary validation, a validation strangely posited in an acknowledged and scientized falsity, will become a common recurrence in this book, particularly on the questions of infrastructure, nation, and economy.

The critique of models, however, partly rendered in this book as it is through a Foucauldian approach to technologies of signification, goes even a step further: it is not only, from the standpoint of scientific methodology, that reaching conclusions through simulation creates the extra step of verifying the accuracy of the simulation itself (whether theoretically or empirically). Additionally, we might conceive more strongly of this process of *verification* (or *validation*, it is sometimes labeled) as, at the same time, a process of *veridiction*. It is axiomatic, conceptually and historically, that if one wishes to use a particular simulation technology to answer a question scientifically, one needs to ensure that the model through which the simulation proceeds is reliable in reporting those facts. But given the tautological dilemma Badiou poses to the model, it is clear that verification has a broader epistemological task. Writing of formal mathematics, he clarifies:

> Now, to be precise, the effectuation of semantic correspondence is *nothing other* than verification itself. It permits us to evaluate the type of scriptural rigour that can be claimed for the domain in question. Control (technical control) of the formal system permits the inscription of a *proof of deducibility* [*déductibilité*] relative to the informal demonstrations that constitute its various models.[58]

In other words, this correspondence between inside and outside, between the system and the so-called empirical reality of the thing in question—or, rather, *the procedures that establish this correspondence*—provides a verification of the truth value of the system itself, proof that the models will produce conclusions both germane and accurate to the questions being asked. To ask after "scriptural rigour" is nothing more than to ask whether the simulating system is accurate, which is also to authorize accurate simulation as a legitimate mode of knowledge production. This is effectively the same tautology as the one identified in structural anthropology, brought to the level of formal logical systems.

But historically, those "informal demonstrations" that are the models, and the simulations they allow for, are more than simply formal proofs for any particular system or model: they are also retroactive proofs for the activity of modeling and simulation itself. That is, in coordinating a set of parameters for themselves, deploying a simulation from within the architecture of the model, and then formulating the outputs of that simulation as a formal verification, they produce the truth value of simulation itself. "Retroactivity" in this register refers to the way in which the lining up of the facts bends backward to authorize the model. Its accuracy will always be ensured by the subsequent success of the representations it produces. Again and again in this book, cases and events under investigation show this sequence. A practical question (usually an engineering one) is broached, a model of some kind is designed and constructed, and simulations are run to prove the veracity of the model, which in turn is understood as a demonstration of the veracity of simulation as a methodological tool.

Given the distortions involved in fitting models to available technologies and ideologies, the circularity of methodological certainty in their simulations, and the strength of critique against the conceptual basis of them, it would be fair to wonder at the purpose of, not to mention the justification for, envisioning these instruments as technologies of public memory. The simplest answer to this question is, quite straightforwardly, that the verification of natural disaster models has been conducted throughout the contemporary era via historical data. In Chapter 3 we'll see that the largest physical model of the early twentieth century, the Mississippi Basin Model, was verified through a simulation of flood stages during the Great Mississippi Flood of 1927. And in Chapter 4, I trace how cartographic models of storms, floods, and hurricanes were generated through statistical modeling methods that imposed a kind of data normalization and transformation of past storms to generate hypothetical ones. And in Chapter 5, I examine the longstanding narrative of The Big One in New Orleans and the manner in which past disasters like 1992's Hurricane Andrew and 1998's Hurricane Georges overdetermined the conceptions of this worst-case storm-to-come and the disaster simulations that immediately preceded Hurricane Katrina. As a matter of method, there is nothing at all controversial about running historical data through a statistical models, whether they are computationally demanding or not. The practical accuracy of that procedure will depend almost entirely on the architecture of the model rather than on the fact that historical data was used, even in cases where the historical record is fairly limited in quantity and/or detail. So, the first reason is: The science of disaster prediction has rested fully upon the enshrinement of past disasters within the very technics of the model.

By itself, this is no more than a description of experimental and/or theoretical method; but when put into the context of the dynamic of public memory raised at the beginning of the previous chapter—the relative absence of public memory objects surrounding natural disasters in spite of the vast scale of human misery and loss of life they cause—the particular scientific-rhetorical forms that disaster simulations have taken throughout the twentieth and twenty-first centuries appear to serve a different function. Emergent as they are from within their political, scientific, and technological milieu, these forms help craft the contemporary understanding of "natural" when speaking about natural disasters. The second justification for treating models as memory, then, is to be able to track the manner of the production of the natural in disaster discourse, and hence the production of the modern form of the signifier "natural disaster." The displacements at work in extracting from disaster discourse the agency of technologization are, I argue, symptomatic of unconscious mechanisms of public memory: as the technologization of settlement and development reaches its point of failure in the face of climatological phenomena, and as that very technologization of the environment contributes directly towards the means to failure (directly through alterations to the landscape or indirectly through anthropogenic climate change), the consolidation of a concept of the "natural disaster" builds a permission structure constituting an uncomplicated necropolitics. In doing so, the mass death and ruination these events cause are laundered rhetorically through the agency of the earth—through the production of the natural as a signifier, that is. It is no surprise, then, that we do not commemorate that for which we disavow responsibility, even and especially to ourselves.

That the proper position of the model is between the consolidation of past knowledge and the projections of the future renders the epistemic practice of modeling a rich index of how historical knowledges of disasters and their prevention and/or anticipation have taken form in pursuit of deliberative ends. This, more than anything, marks the importance of the present inquiry for the study of the rhetoric and politics of the model. The question most centrally at issue in a rhetorical history of disaster modeling is resolutely not the various modeling technologies' capacities for producing useful or effectively actionable knowledge, defined (retroactively) within an ostensibly deliberative field of practice as "correct" or "accurate." Not only does this determination concede important deliberative questions (Useful for whom, and to what end? Why those questions rather than other ones?), it also takes refuge in that supposition of the independent veracity of nature, thus successfully disavowing the question of the political altogether. I invoke Badiou once more to complete the critique:

> The problem is not, and cannot be, that of the representational relations between the model and the concrete, or between the formal and the models. The problem is that of the *history of formalization*. "Model" designates the network [*réseau*]

> traversed by the retroactions and anticipations that weave this history: whether it be designated, in anticipation, as break [*coupure*], or in retrospect, as remaking [*refonte*].[59]

In other words, the question at hand is not whether the model fits accurately enough the reality at a representational level, but, rather, how the model and reality interface in its emergence. What were its features, its forms, its assumptions, its points of contention, its absences? And what do those formal elements tell us about the questions being asked, the problems attempting to be solved, and the rhetorical network of relations that produce the model as a source of truth? The model, often posited too simply as logically and chronologically prior to the act of simulation, must be instead placed within a much more complicated historical sequence of formalization. Each model's history twists forward and backward, its truths existing in both projective and retroactive temporalities.

THE PRODUCTION OF THE NATURAL

In pursuing a rhetorical genealogy of disaster modeling rhetoric in the twentieth-century United States, it will therefore not suffice to describe and analyze the relationship between formalized models and their "natural" counterparts to determine to what degree they have been effective, politically negotiated and acted upon, or even the effects of the knowledge they have produced. These are important questions, to be sure. But, following Badiou, ireading the relationship between artifice, on the one hand, and the nature that it derives laws to describe and simulate, on the other, concedes the duality of artifice and nature far too quickly. To put it succinctly, "natural" disasters are never wholly natural and in the U.S. South are quite often specifically racial, and the accelerating anthropogenic climate crisis renders any conception of the natural as distinct from the artificial incoherent by definition.

Scholars across the humanities and social sciences have for a long time affirmed the basic line of reasoning that challenges the ontological split between natural and human-caused disasters: factors as wide-ranging as infrastructure, insurance law, response resource allocation, personal wealth, and communication technologies make practically irrelevant a distinction based solely on the perceived catalyst, beyond the specific response challenges that climatological events can present. Hurricanes, earthquakes, floods, tornadoes, droughts—insofar as disaster has become a political matter of governance and containment, their occurrence may be "natural," but their effects certainly are not. Plagues decimated Europe for hundreds of years in no small part due to the concentration of bodies and issues of sanitation and health. Disasters are, as I have said, by definition anthropocentric affairs, and historical disasters are written and remembered as such because of the effects they had on human life and its well-being. Moreover, disasters in the modern era, in addition to being assessed on purely climatological metrics, are appraised and remembered for their collective capacity to overwhelm state containment and response, or, in the inverse, for the failures of the state to adequately contain the natural environment. This impulse to assess disaster events at their central and most damning points of failure is the result of centuries of human conduct that has determined in what capacity states, institutions, and individuals are responsible for the factors that determine and delimit those effects. Thus, when those points of failure continually disadvantage and devastate certain groups more than others, it is not unreasonable for critics to question the presumed ontological split. It is just this pattern that causes Robert Mizelle to argue, for instance, that "environmental disasters provide some of the most telling and revealing snapshots of our racial fabric across time, offering an invaluable lens into the meaning of race, citizenship, and the environmental world"[60]—not that they reveal primarily the awesome power of nature. When so many disparities cause massive differences in survivability, resilience, and emotional trauma, the precise cause of the event having been climatological or environmental—that is, privileging a categorization of the origin rather than the effects that are the conditions of possibility for its recognition as a disaster—is not a particularly meaningful observation in the context of political critique. Even worse, scholars argue widely, privileging the frame of the natural has the effect of an enthymematic political absolution: a so-called act of God simply cannot be helped.

The line of criticism is an important one, but, to complicate the matter further, this rationality is not merely a result of the artificiality of the distinction between natural and unnatural; it has to do also with the very concept of disaster itself. Its intelligibility relies upon a centuries-long conceptual articulation of the concept of the natural and the concept of disaster. Disaster, "derived from the Italian *dis-astrato*" and ultimately from the Latin *astrum*, translated as star or constellation, refers originally to a form of cosmic abandonment, a state of having been disavowed by the heavenly bodies and thus left tragically to one's own fate.[61] As early as the post-Enlightenment transition into the modern era, "disaster," as a term, but also a concept heavy with metaphysical import, underwent an historical shift from this theological frame to the more recent sense of the term, what Marie-Hélène Huet refers to as "an administrative problem of political containment."[62] Whether in the context of plagues, earthquakes, or droughts, natural disaster in the modern sense is commonly understood as an interruption in an otherwise normal state of affairs, something as equally uncontrollable as the abandonment of the gods but whose effects need to be contained and responded to. It was this notion of intervention, this constitutive externality outside of human control but impressing upon it, that the concept of the natural merely confirmed. As Huet notes, "the dark side of the Enlightenment itself yielded a world of shadows where dreams of controlling a rebellious nature gave rise to a scientific project itself fraught with perils and anxieties."[63] Thus, from the early beginnings of the post-Revolutionary state, the tentatively secular positing of "natural disaster" hitched itself to the larger project of the art or science of governance, inflecting the specific exigencies of containment associated with managing external, uncontrollable threats to preserve the welfare of populations.

In the United States, the institutional and cultural genealogy of disaster politics has come unsurprisingly to imagine such containment from within the logic of neoliberalism in its many manifestations. In a Foucauldian vein, we might note that over the better part of the past century the responsibility for survival has systematically been offloaded onto the individual. From suburban Cold War-era fallout shelters to contemporary disaster "prepping," the state's role in containing the effects and production of disaster has been eroded in favor of both privatized entities (such as vastly unregulated insurance and surveillance technologies), nongovernmental philanthropic efforts, such as the programs that feed and house disaster victims or rebuild destroyed neighborhoods, and individual responsibility. Denizens of the U.S. are encouraged to ready themselves with supplies, exit plans, shelters, and other contingency planning, that self-entrepreneurialism that Foucault recognized as the proper domain of a new *homo economicus*, a subject who now exists to produce its own satisfaction—in this case, its own survival.[64] Majia Holmer Nadeson explains that "Foucault and subsequent governmentality scholars argue that neoliberalism—as a body of knowledge, strategies and practices of governance—seeks to divest the state of paternalistic responsibility by shifting social, political, and economic 'responsibility' to privatized institutions and economically rationalized 'self-governing' individuals."[65] She uses Foucault's conception of neoliberalism to argue that the moralizing discourse of neoconservatism blamed social programs for the devastation of human lives during Katrina rather than acknowledging their root in neoliberalism more broadly.

For others, the most destructive aspect of contemporary U.S. disaster policy is the way is has unfolded within a hyper-capitalist rationality. As Cedric Johnson argues, "Since the Reagan era, Democrats and Republicans, black, brown, and white elites, and state and federal officials have increasingly converged around the view that private institutions and individuals—not government—should bear primary responsibility for social welfare, education, and disaster preparedness, relief, and reconstruction."[66] Johnson's argument performs no centrist flattening, however: this broad complicity, he contends, spans from the manner in which free market capitalism has crippled social institutions and exacerbated vulnerability, to the outsized valuation of administrative performance *during* disaster crises rather than the mitigation thereof, to the overpriced and easily corruptible private contracting involved in long-term recovery efforts, most often at the expense of the opportunity to put those most affected into jobs that would help their own recovery. Important to his critique is that scholars focusing too heavily on neoconservatism—that is, on a particularly insidious and corrupt manifestation of neoliberalism—fail to diagnose and critique a broader governmentality organizing disaster policy that cuts across major political parties, one whose response entails offering federal subsidies for already exorbitant private recovery contractors rather than a New Deal-style state operation that trains and hires teachers and local business owners who have lost everything to rebuild their own schools and stores.

Still others, like Naomi Klein, see disaster as the precise framework for exploiting trauma and national instability to institute neoliberal state regimes in order to establish flows of capital from state to industry. In her widely read *The Shock Doctrine*, she details the ways in which disaster and instability have been exploited and/or produced by mass privatization, marking out an entire domain of neoliberal economic development designed to craft governmental and other institutional conditions amenable to unrestricted free market economies. There is but one motive of "disaster capitalism," as she terms it: "Far from freeing the market from the state, these political and corporate elites have simply merged, trading favors to secure the right to appropriate precious resources previously held in the public domain."[67] Induced into a kind of "shock treatment," whether by hurricane, coup, or economic collapse, Chicago School-style free market economy oversees the selling off of national industry and labor protections, ensuring that, in the name of economic productivity and development, the proper role of the state is to maintain the free market at all costs.

In these critiques and others, one can begin to detect the broad conceptual and institutional considerations in thinking through the articulation of disaster and neoliberal governance. An argument that insists across them, however, and one familiar to critiques of neoliberalism in its many forms, is the rejection of the purity of market relations as a rationale for the systematic decommissioning of state responsibility for disaster containment. As Foucault details in his 1978–1979 lectures, classical liberalism as an art of government was most chiefly concerned with mapping the effects of governance such that less of it could be done, ensuring individual liberty and freedom. Liberalism, he suggests, may have in any case been a misnomer, for the assurance of liberty and freedom was not, in fact, the condition for the limitation of government: "what we see appearing in the middle of the eighteenth century really is a naturalism much more than a liberalism."[68] It was the observation and systematization of the effects of government, the determination of a political economy, that chalked out the limits of liberty and freedom, and its field of effects was understood to be perfectly natural in that there were natural laws that determined them. By the time of the rise of neo-liberalism, economic freedom became the founding principle of state limitation, and the analytic task of political economy—that is, the practice of determining the effects of certain policies and laws and other mechanisms of government in order to limit it—gave over fully to the pretensions of the free market, which was presumed to operate under its own natural laws so long as the state didn't pollute it with undue regulation. The role of the state under neoliberalism becomes clear: "a state under the supervision of the market rather than a market supervised by the state."[69] The larger argument in which each of the previous criticisms of neoliberal disaster policy participates, then, is not merely that personal and state infrastructure to anticipate, resist, respond to, and recover from disasters is often inadequate and disparate; to the contrary, in deploying the resources of the state to maintain the space of economic freedom, U.S. disaster policy *produces* disaster capitalism as the principle of limitation on which neoliberalism is founded. In other words, these policies are working precisely as designed, and the popular narrative that the destruction of state and federal disaster containment capabilities represents a failure—as was seen in post-Katrina New Orleans and post-Maria Puerto Rico—is not much more than a disavowal of that fact.

It is particularly important to take the critique to the level of governmentality—to the emergence and maintenance of neoliberalism—for two related reasons. First, this book primarily focuses on federal scientific discourse, specifically on the work and influence of the Army Corps of Engineers. As a limitation, following the scientific narration of the state almost exclusively is perhaps not the most comprehensive method of discerning a political rationality, even if those institutions are at the fore of the epistemological technologies within the domain of disaster science. The Foucauldian portion of the method is not merely compensatory for the limited object domain of the book's archive, however. Over the course of critique, we will also be able to detect in this scientific rhetoric the subtle and important shift in the positioning of disaster engineering, the state's perceived role in disaster, its capabilities and responsibilities, and—for, it can never be forgotten, the central question is not of responsibility but of the threshold of acceptable risk—what level of death can be reasonably tolerated. From the massive public works of the first half of the twentieth century, to the crumbling levee systems of New Orleans in the first years of the twenty first century, the science of prediction is always married to the framing of government disaster science *vis à vis* political rationalities.

Suturing disaster studies and the political philosophical critiques of the natural makes clear not only that, for one, disasters have never been natural in a strong sense of the term. It is also, and this is the third and final gambit of this project, that a thoroughgoing critique of the *rhetorical production* of the natural functions as an opening into an investigation of how the collective fantasy of the natural and of the natural disaster impinges on the collective memory of disaster in an era in which we know that even so-called natural disasters have "unnatural" origins—factors which not only have not been solved by the promises of neoliberal capitalism, but which have in fact been accelerating precisely by the political actions neoliberal doctrine sanctions. It is very difficult to imagine what story a collective discourse of public memory might tell in the face of such layered repressions and displacements. And the issue of allowable risk, inextricable from the perception of federal responsibility for disaster and the statistical management of life and death, bears directly on the perception of how to remember and what to mourn. But in the end, at the level of form and formalization, tracing scientific rationalities symptomatically across time and in relation to prevailing philosophies of disaster management reveals collective memory to transfer through the science itself, memorializing events long forgotten in the very architecture of the science, in its methods, its theories, its impulses, and its data.

* * *

The decision to stitch together these three domains—public memory studies, the rhetoric of scientific modeling, and the rhetorical production of the natural—was arrived at somewhat inductively through the course of research, as cobbled as it may appear to some readers to be. To begin an inquiry with first and foremost a question—what is the public memory of natural disasters?—is to suspend the presumed architecture of the framework until the argument begins to form from within the discourse itself. To answer that question, I could not turn to public memory studies because I lacked both epideictic rhetoric and a substantial or particularly insightful archive of place-centered memory objects. So, I first turned to Freud to help read discourses of memory where they are not supposed to be, and, in doing so, I came upon the question of models again and again: their accuracy, their utility, and their history, their epistemology, and most of all their predictive capacity. It was then only a matter of wondering how, over time, these models impacted how disasters were interpreted (not to mention their direct influence on the severity of the effects), and transliterating, as it were, the concept of interpretation into the language of memory. The effect of this movement was to find always the fantasy of the natural at the heart of the matter, as well as the question of state responsibility, which of course leads inexorably to a discussion of biopolitics and necropolitics. Tracing disaster modeling's movements through time gave the book valence, even if the aim is not necessarily to tell a comprehensive tale about the history of such modeling.

The following chapters assemble the framework through this general history by pausing on the most dominant and impactful forms of disaster modeling, mostly though not exclusively implemented by the U.S. Army Corps of Engineers. (Unsurprisingly, many of the discourses moving into the late-twentieth and early-twenty-first centuries arrive increasingly via government contracts and subsidized university research.) Each chapter treats one or two of these forms, reads the science through a Freudian rhetorical lens, and connects the formalization of that science to the broader discussions of disaster happening at the time. In so doing, we will watch not just the form, but, even more interestingly, the *function* of modeling transform along with the rise of late-twentieth century and early-twenty-first century political rationalities. Ostensibly, the most basic function of the model is to be able to predict: the severity, the effects, and the general capabilities for response. But a science of prediction, bounded by the questions of scientists and isolated from other questions intentionally, cannot but ferry the past, through the present, and into the future.

NOTES

1. Pierre Nora, ed., *Realms of Memory: Rethinking the French Past*, vol. I: Conflicts and Divisions (New York: Columbia University Press, 1996), 17.
2. Nora, *Realms of Memory*, xvii–xviii.
3. Nora, *Realms of Memory*, 3.
4. See, for instance: Joseph L. Llobera, "Halbwachs, Nora and 'History' Versus 'Collective Memory': A Research Note," *Durkheimian Studies* 1 (1995): 42; and Paul Ricoeur, *Memory, History, Forgetting* (Chicago: The University of Chicago Press: 2004), 91.
5. Patrick H. Hutton, *History as an Art of Memory* (Hanover, NH: University Press of New England, 1993), 10.
6. Kendall R. Phillips, "Introduction," in *Framing Public Memory*, ed. Kendall R. Phillips (Tuscaloosa: University of Alabama Press, 2004), 2. This tension between history and memory has also spawned what might be seen as hybrid approaches, like Thomas Dunn's, that have an "openness to admitting anachronism while *simultaneously* seeking resonances between the past and our contemporary moment." Thomas R. Dunn, "'The Quare in the Square': Queer Memory, Sensibilities, and Oscar Wilde," *Quarterly Journal of Speech* 100, no. 2 (2014): 219.
7. Aristotle, "De Memoria et Reminiscentia," in *The Basic Works of Aristotle*, ed. Richard McKeon (New York: Random House, 1941), 608.
8. Kendall R. Phillips, "The Failure of Memory: Reflections on Rhetoric and Public Remembrance," *Western Journal of Communication* 74, no. 2 (2010): 214.

9. Aristotle, "De Memoria," 610–11.

10. Aristotle, "De Memoria," 608.

11. Phillips, "The Failure of Memory," 210.

12. Phillips, "The Failure of Memory," 210.

13. Carole Blair, Greg Dickinson, and Brian L. Ott, "Introduction: Rhetoric/Memory/Place," in *Places of Public Memory: The Rhetoric of Museums and Memorials*, eds. Greg Dickinson, Carole Blair, and Brian L. Ott (Tuscaloosa: University of Alabama Press, 2010), 7.

14. Blair, Dickinson, and Ott, "Introduction: Rhetoric/Memory/Place," 27–28.

15. Michel Foucault, *The Archaeology of Knowledge & the Discourse on Language*, trans. A. M. Sheridan Smith (New York: Pantheon, 1972), 120.

16. Foucault, *The Archaeology of Knowledge*, 119; it should also be noted here that "enunciation" for Foucault is of a different order than for Lacan, whose usage of the term is detailed in the paragraphs that follow.

17. Joan Copjec, *Read My Desire: Lacan Against the Historicists* (London: Verso, 2015/1994), 7, emphasis in original.

18. Copjec, *Read My Desire*, 6–7, emphasis in original.

19. Copjec, *Read My Desire*, 8.

20. Copjec, *Read My Desire*, 9.

21. Copjec, *Read My Desire*, 9, emphasis in original.

22. Sigmund Freud, *The Interpretation of Dreams*, trans. James Stranchey (New York: Basic Books, 1955), 295, emphasis in original.

23. Freud, *The Interpretation of Dreams*, 324.

24. Freud, *The Interpretation of Dreams*, 302.

25. Freud, *The Interpretation of Dreams*, 301.

26. "Ces pourfendeurs de Freud ignorant que, loin de réduire la vie psychique aux organes qui accomplissent l'acte sexuel, la psychanalyse entend si et comment l'excitation, la douleur ou le plaisir s'intègrent dans l'architecture complexe des sensations, paroles, pensées, projets." Julia Kristeva, *Pulsions Du Temps* (Paris: Fayard, 2013), 146, my translation.

27. Christian Lundberg, *Lacan in Public: Psychoanalysis and the Science of Rhetoric* (Tuscaloosa, AL: The University of Alabama Press, 2012), 69–70.

28. Although dreams and speech may share tropologies, however heterogeneous descriptions of such a tropology may be, the idea of a collective unconscious is somewhat riskier to posit. Freud seems to dismiss the idea of it—but only under the pretense that the content of the unconscious is collective to begin with: "In what psychological form the past existed during its period of darkness we cannot as yet tell. It is not east to translate the concepts of individual psychology into mass psychology, and I do not think that much is to be gained by introducing the concept of a 'collective' unconscious—the content of the unconscious is collective anyhow, a general possession of mankind." Sigmund Freud, *Moses and Monotheism*, trans. Katherine Jones (New York: Vintage Books, 1939), 170.

29. Jacques Lacan, *Écrits*, trans. Bruce Fink (New York: W.W. Norton & Company, 2006), 677, translation modified.

30. Lacan, *Écrits*, 758.

31. Claire Sisco King, "Rogue Waves, Remakes, and Resurrections: Allegorical Displacement and Screen Memory in Poseidon," *Quarterly Journal of Speech* 94, no. 4 (2008): 430–54.

32. Joshua Gunn, "Hystericizing Huey: Emotional Appeals, Desire, and the Psychodynamics of Demagoguery," *Western Journal of Communication* 71, no. 1 (2007): 1–27.

33. This sort of memory critique remains valuable, of course, but also remains limited to theorizing the rhetoric of what Julia Kristeva has called the sacred and, thus, enact a "theologization of the thetic"—in other words, a continued faith in the analysis of memory's positivities. See Julia Kristeva, *Revolution in Poetic Language*, trans. Margaret Waller (New York: Columbia University Press, 1974), 78.

34. Bradford Vivian, *Public Forgetting: The Rhetoric and Politics of Beginning Again* (University Park: Penn State University Press, 2010), 29.

35. Indeed, Simon Stow's examination of remembering and forgetting in Bush's "Jazz Funeral" speech takes exactly the approach that Vivian critiques. By deploying a most shallow and touristic interpretation of these ceremonies, Stow argues, Bush capitalizes on the grieving process to enact a forgetting of the systemic causes of the storm in order to advance his own policies. Simon Stow, "Do You Know What It Means to Miss New Orleans? George Bush, the Jazz Funeral, and the Politics of Memory," *Theory & Event* 11, no. 1 (2008).

36. King, "Rogue Waves," 431.

37. King, "Rogue Waves," 446.

38. King, "Rogue Waves," 448.

39. See also Marita Sturken, *Tangled Memories: The Vietnam War, the AIDS Epidemic, and the Politics of Remembering* (Berkeley, CA: University of California Press, 1997), 44–84. Sturken's principal focus in this chapter is on memorialization practices and remembering at the Vietnam Veterans Memorial in Washington, D.C. and concerns the careful screening of the traumas of Vietnam.

40. Pierre Nora, "Preface to the English-Language Edition," in *Realms of Memory: Rethinking the French Past,* vol. I: Conflicts and Divisions (New York: Columbia University Press, 1996), xvii–xviii.

41. Nora, "Preface," xvii.

42. Nora, "Preface," xxi.

43. Sigmund Freud, *The Interpretation of Dreams*, trans. James Stranchey (New York: Basic Books, 1955), 442.

44. Given the aforegoing discussion, the reference should be clear: Michel Foucault, *The Order of Things* (New York: Vintage Books, 1970).

45. Aimee Roundtree, *Computer Simulation, Rhetoric, and the Scientific Imagination: How Virtual Evidence Shapes Science in the Making and in the News* (Lanham, MD: Lexington Books, 2013), 3.

46. Jean Baudrillard, *Simulacra and Simulation*, trans. Sheila Faria Glaser (Ann Arbor, MI: The University of Michigan Press, 1994), 16.

47. Baudrillard, *Simulacra and Simulation*, 1.

48. Famously, the May, 1968 uprising inaugurated on the precise day that Badiou was slated to deliver the second half of this talk. He never finished the talk, though

it exists now in print. He credits the uprising with shifting his philosophical program completely: "I pass from a structural materialism, which privileges the letter (the mark), to an ontological materialism, which privileges the evidence of the 'there is,' in the form of pure multiplicity." Alain Badiou, *The Concept of Model: An Introduction to the Materialist Epistemology of Mathematics*, ed. and trans. Zachary Luke Fraser and Tzuchien Tho (Melbourne, Australia: re.press, 2007), xxxvi.

49. There were many French theorists in the late 1960s and early 1970s who took from Russian formalism and its French popularizers some rendition of the distinction between the semantic and the syntactic. It is straightforward enough in retrospect to note, without feeling pressed to pledge formal allegiance one framework or another, that Badiou like so many others is attempting to separate epistemologies that proceed deductively—often tautologically—from their own self-imposed truth parameters from those that simultaneously seek to theorize their own ontological emergence, or formalization, as he would later have it.

50. Badiou, *The Concept of Model*, 17.

51. Badiou, *The Concept of Model*, 17.

52. Sandra Harding, *Whose Science? Whose Knowledge?: Thinking from Women's Lives* (Ithaca, NY: Cornell University Press, 1991), 2.

53. Jacques Lacan, "Seminar on 'The Purloined Letter,'" in *Écrits*, trans. Bruce Fink (New York: W. W. Norton & Company, 2006), 30.

54. Slavoj Žižek, *Enjoy Your Symptom!: Jacques Lacan in Hollywood and Out* (New York: Routledge, 2008), 13.

55. See, for instance, to list only a fraction: David Beer, "The Biopolitics of Biometrics: An Interview with Btihaj Ajana," *Theory, Culture & Society* 31, nos. 7–8 (2014): 329–36; Kevin Brock, *Rhetorical Code Studies: Discovering Arguments in and around Code* (Ann Arbor: University of Michigan Press, 2019); Lauren E. Cagle, "Surveilling Strangers: The Disciplinary Biopower of Digital Genre Assemblages," *Computers & Composition* 52 (2019): 67–78; Celeste M. Condit, John Lynch, and Emily Winderman, "Recent Rhetorical Studies in Public Understanding of Science: Multiple Purposed and Strengths," *Public Understanding of Science* 21, no. 4 (2012): 386–400; Steve Fuller and James H. Collier, *Philosophy, Rhetoric, and the End of Knowledge: A New Beginning for Science and Technology Studies* (New York: Routledge, 2004); Harding, *Whose Science?*; Andrew Iliadis, "Algorithms, Ontology, and Social Progress," *Global Media and Communication* 14, no. 2 (2018): 219–30; Steve Jones, "People, Things, Memory and Human-Machine Communication," *International Journal of Media & Cultural Politics* 10, no. 3 (2014): 245–58; Sarah C. Nelson, Joon-Ho Yu, and Leah Ceccarelli, "How Metaphors About the Genome Constrain CRISPR Metaphors: Separating the 'Text' from Its 'Editor,'" *The American Journal of Bioethics* 15, no. 12 (2015): 60–62; Roundtree, *Computer Simulation*; Michelle Wilson, "Raising the Ideal Child?: Algorithms, Quantification and Prediction," *Media, Culture & Society* 41, no. 5 (2019): 620–36;

56. Roundtree, *Computer Simulation*, 4.

57. Roundtree, *Computer Simulation*, 5.

58. Badiou, *The Concept of Model*, 42, emphasis in original.

59. Badiou, *The Concept of Model*, 54–55.

60. Mizelle, *Backwater Blues*, 24.

61. Marie-Hélène Huet, *The Culture of Disaster* (Chicago: University of Chicago Press, 2012), 3.

62. Huet, *The Culture of Disaster*, 2–3.

63. Huet, *The Culture of Disaster*, 7.

64. Michel Foucault, *The Birth of Biopolitics: Lectures at the Collège de France 1978–1979* (New York: Picador, 2004), 226.

65. Majia Holmer Nadeson, "Hurricane Katrina: Governmentality, Risk, and Responsibility," *Controversia* 6, no. 1 (2008): 72.

66. Cedric Johnson, ed. *The Neoliberal Deluge: Hurricane Katrina, Late Capitalism, and the Remaking of New Orleans* (Minneapolis: University of Minnesota Press, 2011), xxiv.

67. Naomi Klein, *The Shock Doctrine: The Rise of Disaster Capitalism* (New York: Picador, 2008), 18.

68. Foucault, *The Birth of Biopolitics*, 62.

69. Foucault, *The Birth of Biopolitics*, 116.

Chapter 2

Physical Scale Modeling and the Rhetoric of Sublimation

INTRODUCTION

The age of the machine in the United States is the historical point of departure for this study of disaster modeling for a number of reasons. For one, the rise of high modernism and the consolidation of new and important scientific fields marks the emergence of an ideological system that extended from the celestial sciences to theories of the state and economy. At the very same moment that advances in industry and science produced the seemingly infinite potentiality of technological mastery over the natural world, the principles of liberalism that first turned government into an art of "how to not govern too much," as Foucault has summed it,[1] were transposed into the establishment of a managerial state. At a glance this might seem a contradiction: How can the *laissez-faire* directive of economic liberalism guide the establishment of a managerial form of governance wrought from conservationism and the desire, not merely to adhere to the dictates of the natural world but, rather, to conquer nature through technological mastery of its principles? In truth, Foucault's analysis of the art of government makes clear what is today axiomatic to scholars of liberalism, chiefly that the proper activity of *laissez-faire* capitalism is made possible only by enormous state effort to maintain the conditions whereby markets are perceived to be free and to have "natural" principles, rulesets that must remain unimpeded by the meddling of the state to function properly.[2] The managerialism of the machine age was quite simply the application of high modernism's unrivaled faith in science's analytic framework—and its presumed capacity to uncover the secret principles that make the world function in the ways it does—to the question of *both* governance and the natural world. As John Jordan has explained it, the presumption of this application was that "the same methodology that enabled steel to be manufactured to previously unattainable degrees of hardness could solve ethnic tensions or alleviate poverty."[3] Once one discovered those principles, one could produce a system that allowed the variables within it to move freely according to their nature, but with a particular set of administrative aims.

So it was with the natural environment as well, particularly when it came to questions of flood control in the Mississippi Basin. The Mississippi River had long been seen as an administrative problem in need of a solution, and engineering became the primary epistemic frame through which to imagine possible solutions. The seasonal flooding that seemed to be getting more and more destructive every year in the post-Reconstruction era and throughout the 1920s presented itself as an economic imperative, to be sure, but it was primarily understood as a physical problem to be solved by technological discovery and expertise. "Corporate capital, organized research, and applied science combined to demonstrate how humanity could understand and control the natural environment," stresses Jordan. "During a structural transfiguration comparable to the Renaissance or any other cultural earthquake, citizens hailed one shining hero of the moment: the engineer."[4] Engineering as a profession grew out of the articulation of the modernist quest for scientific principles and the practical, administrative need to put those principles into the service of industry and the state. Thus, when it came to early conservation and flood control measures, Richard Mizelle, Jr. writes, leaders on that front believed that "science and new forms of technology made the rational planning of landscapes a realm of endless possibility" but also that decision-making regarding such planning "should be in the hands of individuals most qualified to make those decisions."[5] Who was most qualified and which principles of such rational planning were most appropriate were the issues at the core of a national debate ostensibly concerned with the advancement of the engineering sciences, most notably hydrology and hydraulics. But, as I make clear over the course of this chapter, the debate over facilities, buildings, administrative authority, and scientific methodology is best understood as a debate over the representation of disaster and the rhetorical means by which disaster modeling maintained the illusion that disasters were by any stretch natural in the first place—ironically, by attempting to engineer them out of existence. Hydraulic engineers used physical modeling to re-present disasters as natural phenomena devoid of social consequence, thus sealing off the science of disaster prediction and control from more broadly administrative questions of governance, disaster, and, of course, race.

This chapter traces the emergence of a new formalization of disaster prediction knowledge rendered through the practice of physical scale modeling as an engineering science. Engineers, grappling with the explosion of steel and other novel building technologies, needed solutions to equally novel engineering problems, and one idea for solving these problems was to build smaller, scaled-down versions of proposed structures based on principles of similitude. One could then run load tests on miniature bridges, or measure the flow of water over a tiny spillway, or determine the wake patterns of different ship hulls, determining in the process whether their designs would accomplish the goals of the project in the full-size version. These early integrations of physical models were thus indirectly predictive, not the imaginative forecasts of worst-case disasters that we shall explore in later chapters, but rather simulations that aimed to so precisely extrapolate the fundamental principles underlying hydraulic physics that one could design structures capable of handling what was determined to be the scope of disaster's effects. After all, their accuracy was often verified by data collected from previous disasters. During the late nineteenth and early twentieth centuries, scale modeling in engineering contexts was seen as, at best, a practice of inquiry that could only gesture towards the solutions to very specific engineering problems but which lacked the scientific rigor necessary to derive principles; at worst, it was seen as a misguided and illegitimate kind of toy play that not only could not derive general engineering principles, particularly for problems so complex as flood control along the Mississippi River, but also risked producing inaccurate knowledge for individual structures. From this perspective, model study was little more than a methodological innovation practiced only for the edification of theorists, a rejective version of Badiou's observation that, "for the epistemology of models, science is not a process of practical transformation of the real, but the fabrication of a plausible image."[6]

But by the early interwar years, advocates had made some headway in legitimizing the practice. The Great Mississippi Flood of 1927 propelled the issue of flood control (and therefore the desperate attempt to retain both agricultural acreage and the Black labor that secured its profitability) to the top of the federal government's list of priorities, and in 1930 the U.S. formally broke ground on a hydrology laboratory in Vicksburg, MS whose primary function was to model hydrological structures and landscapes for flood and water control projects around the country. By the 1950s and 1960s, "model studies" were so standard a practice that in public works appropriations hearings, officials in the Army Corps of Engineers (hereafter "Army Corps") rarely failed to organize their rhetorical appeals to spending-averse U.S. senators around assurances that model studies of this or that structure had already been completed.

Arguments over the use of models for the production of physical and engineering principles and arguments over the use of models to solve specific engineering problems each conceived of the legitimacy of modeling on similar conceptual grounds, but in each case their proponents often had very different ends in mind. Academics—mathematical physicists and hydrologists, for example—primarily sought after the principles of relation between models and their prototypes[7] and the laws governing physical properties more generally. Professional and military engineers, on the other hand, seemed to care about such principles only to the precise degree to which they informed the design and construction of particular structures, often, as will be seen, in order to mitigate the effects of disasters and maintain waterway commerce.

This technical sequence, laid out in just the last few pages, is one inscribed more or less directly in the primary discourses under examination as well as the secondary literature assessing the period of high modernism, engineering, mathematical physics, and natural disasters up until about the late 1920s.

But the motive structure of the political and technological containment of disaster along the Mississippi during this time has a fairly obvious kernel: the need both to protect the rich agricultural land of the Delta and to constrain the rapidly north-migrating Black labor (most of it sharecropper) that secured its profitability. The 1927 flood and its aftermath capped off a decade in which nearly a million Black Americans left the rural South headed northward. The displacement that the Great Flood caused for hundreds of thousands and the deaths of countless others—some at the hands of white citizens and soldiers who forced Black men at gunpoint to perform physical labor on the levees to save white neighborhoods and businesses—acted as an accelerant in a trend long coming. As Mizelle writes,

> For African Americans living under the yoke of Jim Crow in the deepest South, the 1927 flood was not only an environmental event but also a racial one. While class and immigrant distinctions are often put aside under the umbrella of whiteness during moments of extreme crisis such as the 1919 Chicago Race Riot or the 1927 flood, these moments have always provided a particular type of window into the meaning of blackness. For black survivors of the 1927 flood, they had to deal with the dangers of both the environmental world and their second-class status as citizens where skin color alone was all the justification needed to be murdered.[8]

Richard Wright captured this dual trauma in his short story about the flood published in 1937, "Silt" (later reprinted as "The Man Who Saw the Flood"), in which a man, a woman, a child, and their cow return to their sharecrop farm, silt and mud caked inside and out, only a plough and a dirty well pump salvageable. At a loss for what to do next—the man wants to leave but owes the white man who owns the farm almost eight hundred dollars—he dreads the prospect of starting over again. Looking up at the landowner, he says, "Ah don wanna make no trouble. But this is jus *too* hard. Ahm worse off now than befo."[9] Under the implicit threat of violence and with the perverse cordiality of servitude, the man ultimately and wearily boards the white man's buggy to retrieve new provisions and repeat the cycle. A great many Black Americans instead left. While the 1927 flood was merely one push in a much larger migratory trend during the Jim Crow era, it was such a significant one that "protecting commerce" became an immediate metonym among others for retaining Black labor, and thus for passing federal flood control legislation. This, in turn, helped to justify the establishment of a national hydraulics facility whose primary concern was to be what is known as "river training." Modeling rivers and structures was as much or more about securing arable land and river transportation in the Delta *to be labored upon* as it was about spirited scientific inquiry and sheer engineering ingenuity.

The attendant translocation of the function of (rhetorical) similitude in the practice of modeling during each of the three phases I've isolated in the ascent of this hydraulic laboratory can therefore be traced through the ascent of model study itself as a rhetorical technology of veridiction during this period. Throughout the congressional hearings and debates about the establishment of a national hydraulic laboratory; the post-1927 flood control hearings held to determine finally the location, jurisdiction, and proper function of the facility; and the construction and implementation of the largest model study in the world; the anxieties that wracked both mathematical physics and engineering, both scientific and political will, should be understood as a consequence of a tension between the desire for pure mathematical and physical re-presentation and the desire to re-present—to reimagine—political and social disaster as a pure consequence of nature. The status of the principle of similitude, a geometrical theorem that heralded dimensional likeness as the warrant for the revelation of the laws of nature, ascended in the administrative consciousness of the state from suspicious and potentially dangerous, to supplementary and/or scientifically self-aggrandizing in the worst senses, to standard operating procedure. As it did so, model study transformed from the often-abstract pedagogical instruments of the university lab into tiny renditions of proposed engineering structures, and then finally into the miniaturization of entire geographies and the subsequent rehearsing of previous disasters within them. Physical disaster modeling in the United States thus ultimately found its apotheosis in the Mississippi Basin Model, a 200-acre hydraulic scale model of its titular river basin from Sioux City, Iowa to Baton Rouge, Louisiana.

What was presented in political discourse as a technical, administrative, and epistemic concern was as much or more about preserving the economic and racial structures of production, and the largescale attempts to extricate those questions from issues of race and labor produced a rhetoric symptomatized by a crisis of representation. The representational crisis of the physical scale model, however, was not that of the American technological sublime, a concept David Nye has persuasively detailed in his examinations of extraordinary feats of twentieth-century engineering construction. According to Nye, these feats of engineering like the Hoover Dam and the Golden Gate Bridge produced the experience of the sublime in the witness, an aesthetic experience traditionally associated with the wonders and terrors of the natural world, but which extraordinary technological achievements made possible during this period. While the Mississippi Basin Model did attract a fair amount of tourism during its operational period, the possibility of instilling in its visitors the experience of technological sublimity was not its most notable rhetorical feature. Rather, the downscaling of the model study sought to re-present the physical world such that disasters were impossible—impossible because they were simulations in the first place, impossible because they were calibrated by and engineered to withstand the assumed parameters of disaster, and impossible because they removed both the social consequences of failure and the social conditions necessary for failure to appear as such. These later model studies reproduced the world through a peculiar conception of likeness, always grasping for its philosophical *and* methodological warrant, freezing that likeness at the precise threshold of disaster but going no further. For all their sophistication, abstract laboratory apparatuses could not possibly recreate disaster; the physical forms they took looked nothing like the ravaged worlds and lives out of which their existence emerged. It is no surprise, then, that the greatest physical model study ever performed was the symptomatic attempt to wrest control of the Mississippi from nature by building it anew, absent the question of race located at the center of its understanding of disaster. What hydraulic engineering sought during the first three decades of the twentieth century was not the technological sublime, no, which attempted through grander and grander physical exclamations of concrete and steel to achieve the feeling of the infinite and the mastery of the physical world through the expansion of scale. Rather, hydraulic engineers articulated the achievement of mastery to a different scaling, which was both a down-scaling and a de-scaling, both miniaturization and abstraction. The Mississippi Basin Model thus epitomizes not an aesthetic of the technological sublime, but rather the rhetoric of sublimation: the positive function of a re-presentation, a believable likeness, which in its very utterance signals the persistence of that trauma which it displaces.

COEFFICIENTS OF SIMILITUDE

In all manner of activities, from physics to the art of government, the high modernist period was concerned chiefly with unlocking the secret laws of that which was understood to be natural. Lewis Mumford, progenitor of wide-ranging debates regarding the ascendance of the technics of the machine during this period, sums up the exigence quite simply: "the mass of mankind [*sic*] learned, at least during the period of the written record, that certain parts of the environment can neither be intimidated nor cajoled. To control them, one must learn the laws of their behavior, instead of petulantly imposing one's own wishes."[10] Inherited and transposed from a religioscientific tradition that sacralized nature as itself the Word,[11] Mumford's historical imprecision here is padded by the reader's knowledge that he is referring specifically to the articulation of a post-Enlightenment modernist intellectualism and the height of the machine age. The scientific imperative of the rising field of professional engineering, particularly in the United States, was nevertheless wholeheartedly pragmatic. The metonymic relation in the final sentence of Mumford's quote gives face to this truth: the obvious and apparent function of scientific discovery is control. If one can define the hydrological principles that determine how water moves over a weir-like structure, one will be able to specify more precisely how water will move over *that* weir and render a technical design to direct and constrain the water. If one can determine the exact effects that river cutoffs have on downstream flood stages, then one can anticipate whether a proposed cutoff poses a grave risk of exacerbating floods downstream and engineer accordingly.[12] The major debate in the early twentieth century, however, was what the proper methodology for arriving at those principles should be.

In 1913, esteemed engineer John R. Freeman visited several hydraulic engineering laboratories throughout Europe, including one in Dresden, Germany. These laboratories, some considerably more sophisticated than others, contained structures, devices, and instruments whose purpose was to simulate hydraulic phenomena in a controlled and measurable fashion. Freeman was so taken by this methodology that he became a keen advocate of it. It quickly proliferated throughout Europe during the early 1920s in both colleges and governmental facilities, but its adoption in the United States during that same time was limited primarily to universities. Responding to his and others' advocacy, the American Society of Engineers soon established a traveling fellowship to study abroad at these European facilities.[13] A core irony of the push for the importation of physical hydraulic modeling methods from Europe is that, nearly thirty years prior, Hubert Engels, the founder of the Dresden laboratory, became intrigued by the method while visiting a hydraulic laboratory at the University of Michigan. This somewhat rudimentary laboratory was designed exclusively for the classroom, which is to say, for purely instructional reasons. Ben H. Fatherree writes, "[Engels] was particularly impressed by use of a glass-sided flume designed to illustrate the flow of water over a weir."[14] Engels's fascination was with this novel methodology: beyond equations, graphs, field trips, and other pedagogical technologies of mathematical and hydraulic physics, the University of Michigan laboratory had constructed physical models of various artificial hydraulic structures. The visibility afforded by the glass side marks their pedagogical purpose, as does one other key aspect: the models housed in universities both in the United States and abroad were not, by and large, simulations of actually existing structures, natural or engineered. They were primarily scaled abstractions that demonstrated hydraulic principles for academic and educational reasons. The pragmatic challenge for both pedagogy and scientific analysis was, of course, whether one could trust a scale model to provide correct scientific analysis of a phenomenon occurring at either a larger or completely abstracted scale.

On that question, Herbert Vogel, another influential force behind their eventual adoption, reflected three decades later:

> In 1930 . . . most hydraulic engineers thought of models as mere toys for the youngsters of the profession. A very large group was openly skeptical and many voiced their disdain. This attitude was shared by many officers and civilian employees of the [Army] Corps, and a few held ill-concealed resentment of the newfangled approach. Some high officers actually feared that model testing imposed a threat to sound engineering based on theory and computations.[15]

Off the bat, this historical recounting is simply untrue, as we will see later: a strong majority of civil engineering organizations and individuals supported model testing by the early 1920s[16] (including a hearty endorsement by engineer and then-Secretary of Commerce Herbert Hoover[17]) and an even more vocal show of support appears in congressional records in the years immediately following the 1927 flood.[18] It was the Army Corps itself that stood practically alone in opposition to the methodology. Nevertheless, the sequence is fairly accurate, even if the dates are conveniently wrong: there persisted for some time among military and civil engineers alike the prevailing belief that model testing lacked a sound basis for producing adequate theoretical knowledge—not that it was necessarily impractical for solving engineering problems, but that it could not be used as a basis for deriving theoretical principles. Modeling, for some, threatened established theoretical axioms, potentially resulting in disastrous outcomes that would be otherwise preventable by sound theory.

The prevailing conceptual rationale for scale modeling, and thus the explicit rejoinder to this question, has for a long time been the principle of similitude. The preliminary deployment of this principle for the purposes of engineering was essentially geometric: "Similitude held that if two geometric figures were drawn so that all corresponding lines were of the same ratio and their corresponding angles were of the same degree, the two figures were similar, and all the properties of one could be determined from the other."[19] In

Figure 2.1. Students from L'École Polytechnique visiting Waterways Experiment Station on August 30, 1948 see methods of studying surface current action in river. *Source*: The National Archives 2, Still Picture Branch, SC 308714.

other words, to take a grade school example, it is one thing to note that a right triangle whose non-hypotenuse sides are 3 and 4 will have a hypotenuse of 5. This information teaches us only about the dimensions of this triangle, and one could discover these empirically if one wished, for instance, by measuring them with a ruler. What Euclidean geometry extrapolates from this is a theoretical principle that remains true no matter the scale: $a^2+b^2=c^2$. The Pythagorean Theorem describes this particular triangle, but also a triangle whose sides are 6, 8 and 10, or 300, 400, and 500. These abstractions are, of course, the bread and butter of the theoretical form of mathematics and, isolated within that theoretical form, they suffice to justify the accuracy of scaling. As long as the proportions remain true, the dimensions matter not. So it was with early notions of physical modeling for engineering purposes, one example of which can be seen in Figure 2.1: obtain the ideal proportions, and one can shrink or enlarge those dimensions with confidence in the behavior of the full-size structure.[20]

These Euclidean model demonstrations were in some sense abstractions not unlike the ideation that Newtonian physics itself made possible through the syntax of its mathematical modeling. As Edwin Layton notes, "At one level, the fact that mathematical models based upon Newtonian science were idealized was an advantage; it made for simplicity and ease of calculation for approximate results."[21] (Badiou, ribbing Lévi-Strauss, jokes in *The Concept of Model* that "the classic condition of elegance" in the construction of models is simplicity.[22]) The syntax of Newtonian coefficients encourages just that. The Newtonian gambit is, as Ivar Ekeland sums it:

> If the positions the planets occupy today are known, together with their velocities, it is possible by these methods to compute their positions and velocities at any future date. It is also possible to turn back the clock, and to find out the positions and velocities of the planets at any past date. Astronomers predict the eclipses of the sun and the moon, and they also figure out the exact dates of ancient eclipses recorded in historical documents. In other words, the past and the future of the solar system are entirely contained in its present. To know the state of the universe at any past or future date (the mathematics sees no difference), it is sufficient to know its present state accurately enough and to have adequate computing power.[23]

However, he also argues, ever since Einstein the field of physics has been forced to reckon with the fact that "Newtonian physics stand today as a very useful approximation, very precise in its own range of applications, but devoid of internal justification, a purely mathematical model."[24] Rendering complex physical processes into a mathematical language involved a formalization of that language, the semantic and syntactic consolidation of its variables into generalizable relationships written into differential equations. As Martin Reuss's work evidences, the problems associated with flooding and river training were no exception to this:

> If permanent solutions were to be found, thought many turn-of-the-century scientists, mathematics would provide them. In a deterministic world, nature's laws were both knowable and computable. Differential equations provided the key. Once integrated, they revealed the trajectory of substances as big as a galaxy or as small as an atom. They extended the empirical approach and heightened the researcher's predictive power.[25]

Ideally these deterministic relationships, primarily applied to celestial bodies at the time, could be extrapolated infinitely into the future or past, leading to such advancements as the backwards dating of ancient eclipses or the prediction (or near prediction) of future ones. If one knew the rules of those bodies, one could apply any temporal or positional inputs to their equations that one wished:

> The most perfect mathematical expression of determinism is the differential equation. It is a mathematical tool which was created by Newton to derive Kepler's laws from the gravitational pull. Even today, a deterministic system is a system which can be modeled by a differential equation. It has the fundamental property that its state at any time completely determines all subsequent (and preceding) states, and there are numerical procedures to compute the state at any time *t* from the initial state and the equation.[26]

Like coefficients, the laboratory instruments found in Dresden, the University of Michigan, and a number of other colleges in the United States and Europe were designed for the elegant and simplistic exposition of complicated principles, mostly for educational purposes. Not every physical process is as simple to render in such observably reliable equations as celestial bodies, however—or, at least, the deviations of planets and stars from established calculations play out over such enormous periods of time that it appears that way relative to the life of theory. But beyond the problematic of mere physical complexity, which will rear its head again later in the chapter, one problem with Newtonian physics was that it had little capacity to account for scale and dimension. This created problems both theoretical and practical.

On a theoretical level, that Newtonian physics lacks a formal conception of scale threatened outright the legitimacy of similitude and thus the entire enterprise of physical modeling. As Layton emphasizes, "a consistent mathematical model of physical reality must incorporate rules to ensure independence of scale, since otherwise the mathematical model derived from physical theory might apply at one scale, but not all scales" and would thus lack general validity.[27] As the mounting complexity and scale of technological projects across the world exposed the limits of these mathematical models to predict how machines and structures (not to mention landscapes) would behave under different conditions, the use of scaled physical models emerged as a potential solution, one that sought both to solve practical problems of engineering and design and also theoretical ones. That is, physical models were also conceived to help formulate theretofore unrealized theoretical principles that could not be tested (or would be too expensive and/or complicated to test) at prototype scale. Thus, the imperative to formalize the principles linking shape, force, and time into rules for building physical scale models whose dimensions could be verified through historical data emerged as the central problematic that could justify the practice of modeling itself. This is the circularity that Badiou critiques unequivocally in structural empiricism's approach to the construction of models but which also takes shape outside of that tradition, *per se*: the model is that which accurately explains all the facts, and the true model is to be verified in accordance with the observed facts. The theoretical exigence was thus precisely to formulate a generalizable symbolic expression of physical processes, to unlock through formalization the secrets of nature and thus the control of it through engineering. The means by which that could be achieved was, of course, by producing a believable likeness of it.

The need to define the parameters of the legitimacy of that likeness was a primary surface of emergence for "dimensional analysis," which attempted above all else to locate the exact relations of dimensional scaling. What is meant by this: relations of scaling? This concept, given a number of different informal terminologies by engineers and scholars at the time, secured the necessary departure from Euclidean similitude. Thomas Wright lays it out clearly: "The justification for using models to predict full sized behavior rested, and still does, upon the idea of the existence of scaling relationships linking physical quantities measured on the model to corresponding quantities measured on the full size."[28] On one level, this is, broadly speaking, the same principle of nondimensionalization as just detailed: a mathematical theoretical practice that replaces dimensions—for instance, the actual length of the sides of a triangle, 3, 4, and 5—with nondimensional variables—*a*, *b*, and *c*. Once you have the equation, you can scale it however you like. But the positing of a scaling *relationship* signals something else, namely, that in any given theoretical formulation, the alteration of scale also relies on one or more *extra* variables or coefficients that attempt to make up for the fact that the physical laws governing one size and those governing another are not, within a certain conceptualization of what those laws describe, the same. In 1915, Lord Rayleigh, in advocating for the principle, laid out a handful of examples:

> Geometrical similarity being presupposed here as always, how does the strength of a bridge depend upon the linear dimension and the force of gravity? In order to entail the same strains, the force of gravity must be inversely as the linear dimension. Under a given gravity the larger structure is the weaker.
>
> The velocity of propagation of periodic waves on the surface of deep water is as the square root of the wave-length.
>
> The periodic time of liquid vibration under gravity in a deep cylindrical vessel of any section is as the square root of the linear dimension.
>
> The periodic time of a tuning-fork, or of a Helmholtz resonator, is directly as the linear dimension.
>
> The intensity of light scattered in an otherwise uniform medium from a small particle of different refractive index is inversely as the fourth power of the wave-length.[29]

The enumeration goes on in the text, but the point has been made: in each of these cases, the task at hand is to identify the specific *extra* calculations necessary in a physical relationship to scale things up or down. Whether inverse linear, direct linear, square root of a particular variable, or "the fourth power of the wave-length," the theoretical component of physical scaling involved isolating and formalizing a portion of the calculation that modifies the dimensional aspects to account for scaling itself in non-static physical contexts.[30]

Part of the rhetorical translation of Euclidean and Newtonian principles of fundamental similitude into physical model studies, then, involved retaining the mathematical foundations of classical geometry and physics, but supplementing them with an additional element—one that was not simply geometrically nondimensional. Proportion was only one part of it. This additional nondimensional element was deployed to "fix" the problems that such equations ran into at different scales, rendering their calculations more accurate to empirically observed conditions in the world. This basic computational structure remains broadly true in dimensional analysis today because it still helps answer perfectly normal questions related to physical scale, for instance, how water moves through variously sized pipes or whether a very tall building will fall down. Thus, the modification of the principles of similitude and the introduction of dimensional analysis must be seen not simply as the continuous innovation of theoretical knowledge about mathematics and physics, but rather as evental in the Badiouian sense: a breaking up of one formalization into another.[31] Referring to slightly earlier shifts in European practice, Wright sums it up neatly: "It will be clear that as theory moved from [geometric or proportional understandings of scaling to mathematical physics], the mathematical apparatus moved from proportion and ratio, to equality and mathematical physics; from pre-nineteenth century to mid-nineteenth mathematics; from a position requiring minimal physical mechanism for understanding to one entirely dependent upon such an explanation."[32] In short, the shift of knowledge formations into the age of the machine made all but inevitable the rise of physical scale models, which not only helped solve individual engineering problems, but also worked to formulate the theoretical supplements to classical principles. This has become, as Margaret Morrison has noted, a fundamental part of mathematical modeling of physical processes: the understanding of many unrepresentable phenomena requires these seemingly artificial supplements.[33]

The conceptual task of that moment, however, was how one ensures that the rigorous theoretical conditions developed in the laboratory will accurately predict what will happen in "nature" at full size, and thus that these coefficients of relation are sound. In short, how can the scientist and the engineer be sure that the representation does not mislead? The rhetorical *stasis*, the point of friction and the center of early debates, was therefore a debate over the scientific process itself. This argument proceeded principally between the Army Corps and the newly professionalized societies of engineering science and practice and animated the political struggle over the establishment of a national hydraulic laboratory throughout the 1920s, setting in place the terms of a debate both philosophical and administrative, and routed directly through the terrains of scientific inquiry and political jurisdiction.

The need to identify those coefficients was the core of Freeman's argument as to the necessity of a national hydraulic laboratory because he believed that model testing was the means to do so. The first congressional hearing before a subcommittee of the Committee on Commerce regarding a joint resolution to establish this laboratory in Washington, D.C. took place in September, 1922. Freeman's testimony in front of the subcommittee takes up the majority of the hearing, and in it he lays out a comprehensive rationale for not only why the laboratory should be built, but also why it should be located in the Bureau of Standards in D.C. While Freeman had been advocating for such a facility for some time, the flooding of the Mississippi in April 1922, in which the highest recorded flood stage in New Orleans reached 21.27 feet,[34] caused many to question the prevailing hydraulic philosophies that had to that point justified the standard practices of river training.[35] After visiting the region of his own accord, he concluded, "now is the time to found this laboratory while memories of the recent terrible experiences are fresh."[36] He makes clear that "tests with models in the laboratory are proposed to be used hand in hand with observations in the field or river, 'on the full-sized specimen,' as a means of obtaining the underlying scientific fact or law of nature with greater precision."[37] The assurance that fieldwork would supplement model study stems undoubtedly from the common critique that laboratory work would have a strictly theoretical or philosophical purpose, thereby neglecting the practical application of such knowledge, but Freeman nonetheless firmly positions model study as a scientific activity in the pursuit of nature's secret laws. When asked later in the hearing, for instance, whether the "fundamentals" of hydraulics are understood, he retorts, "I think they are hardly understood at all."[38] The whole enterprise of laboratory study, for him, led inexorably towards that ideal. Indeed, in a speech to the American Society of Civil Engineers in New Orleans the following year, he expressed confidence that "coefficients of relation between model and full-size original will soon be learned, and after the variants have been tried out on the model, one can go ahead with great confidence in the field."[39]

It is this relationship between model and nature that the Mississippi River Commission's (MRC) representative testimony at the same hearing, made by J. A. Ockerson "at the [insistence] of the Chief of Engineers, United States Army," calls into question.[40] Ockerson asserts that after decades of Army Corps work "in nature's own laboratory, the river itself, it is believed that the commission has accumulated a volume of data covering practically all of the varying phases of the physics of the regulation and control of the Mississippi River" sufficient to control it. It is not merely that the work has been done, however. Ockerson proceeds to lay out a series of ten counterarguments against the establishment of the laboratory, a list equal parts assertive and internally contradictory, but which nevertheless lays the groundwork for a general objection to the practice of modeling itself: the river is too massive and complicated for a model to represent, and in any case the real problem is that levees have broken, not that the physics of the river are too little known. At the level of argumentative content, the entire problem presents itself as an administrative struggle over both method and jurisdiction, and to a very material degree that is a fair characterization. Freeman wanted the laboratory under the purview of the Bureau of Standards, from the standpoint of which theoretical standardization serves as the most obvious warrant; the MRC wished to retain regional control over the study and management of river training, and the obvious warrant would include the primacy of on-site fieldwork.[41]

It would be a grave mistake to conclude the assessment of this debate at the point of this internal philosophical and jurisdictional struggle. As Todd Shallat has gracefully argued, even the Army Corps itself was founded with an internal tension between French philosophies of ordered state planning engineer-scholars and the British and American faith in protestant fieldcraft, and the relationship between civilian engineers and West Pointers was just as often a reversal of that which characterizes this particular debate: "Civilian builders were mostly suspicious of the army's academic training."[42] The distinction between the goals of science and the goals of technology was not nearly as policed in either the Army Corps or civilian ranks as the debate over the hydraulic laboratory seems to indicate.[43] In short, unless one wishes to reduce the laboratory question to a vulgar matter of administrative power, there appears to be very little conceptual or contextual evidence to suggest that a definitive difference of philosophy can be sustained between Freeman's civilian engineer-scientist and Ockerson's Army Corps fieldworker—not even necessarily on the methodological question of the model. Neither position really belongs to the order of truly logico-mathematical discourse and therefore both take on the assumption that even the most theoretically inflected modeling is at root an endorsement of empiricism and the presumption of the object external to the theory. From a *philosophical* standpoint, there is scant daylight between the two.

One obvious symptom of this in the hearings is that both stances in the debate position the failure of the other at the same exact limit point: disaster. For Ockerson, the problem is with the entirety of the laboratory approach. Indeed, he makes the clarification immediately prior to his enumeration of counterarguments that "when I speak of 'models' I speak of the whole scheme of hydraulic laboratory work."[44] Later in the hearing, Senator Joseph E. Ransdell, who helped introduce the joint resolution, asks regarding the recent flooding of the Colorado River and Salton Sea, "Do you imagine that those problems could be aided at all by such a laboratory, Have [*sic*] you thought about that connection?" Ransdell does not seem to have anticipated Ockerson's response:

> That was an emergency. No laboratory could help that. That had to be done at once, and could not wait for experiments, nor would any experiments that might have been made in the past have had any bearing whatever on that situation. And that is the trouble with most of the problems we are up against. They are problems that come suddenly, and we have to meet them at once with what facilities we have.[45]

The implication here is obvious, and involves the entire problematic of the procedures of experimental control measures. Are disasters not in part characterized precisely by the occurrence of an event outside of control conditions, an event that escapes the imaginative boundaries of the possible? For Freeman, the problem with fieldwork is conversely that it is unable to demarcate those boundaries at all:

> The great floods which threaten widespread damage come only about once in 10 years. The Government can not profitably organize an observing staff and fit it out with apparatus or sit down and wait for an opportunity to observe the effect of flood velocities; and when the flood does come, the able engineers of the district will have their hands too full of matters of protection and relief.[46]

Laboratory work, on the other hand, can simulate these conditions, imagine beyond them, and then devise engineering schemes for controlling against the transgressions of disasters. In an incredibly moderated way, Freeman's conception of disaster abjects what Huet recognizes as the tragic dimensions of the premodern era, the propping up of a vengeful god as the agency of disaster, and it also rejects—administratively, if not philosophically—the mere substitution of an uncontrolled nature in its place.[47] That, however, is the extent of the disagreement over the question of disaster. The modern concern with disaster is instead, as we have explored, the necessity of its representation, and to that end the very premise of the debate remains fundamentally intact during the congressional hearings: how does one represent disaster?

On that subject, both Freeman and Ockerson turn to the issue of the water and the land, the wet and the dry, navigation and agriculture: in short, commerce and the production of white wealth in defiance of the dangers of the Mississippi River, the quintessential colonial fetish point in the region. A portion of Freeman's testimony entitled "Dangers that Remain After Levees Are All Completed," for instance, moves seamlessly between the valuation of land and the problems of crevasses, or cracks that develop in levees and violently and suddenly widen, "with the result of putting 5,000 square miles of fertile farm land under water." He first spends a short time summarizing the potential value of this Delta land when properly drained, noting that the current levee system, while expansive and expensive, has not provided "real safety." "By far the most immediate and important use for a national hydraulic laboratory," he then interjects, "would be the development of more precise knowledge about river currents and their hidden method of attack upon a bank supporting a levee and the means by which such an attack can be prevented." This sequence occurs again and again during the congressional debates and hearings between 1922–1930: "30,000 square miles of rich alluvial lands" (or some other estimation) always placed in jeopardy by disaster wrought both by nature and technological failure, followed by the immediate proclamation that "there is no question" that a national hydraulic laboratory would help "to make a comparatively accurate determination in problems of that nature."[48] One letter submitted into the record distills the logic perhaps most efficiently: "Economic reclamation of large areas of fertile land awaits a more scientific knowledge of hydraulics."[49] Disaster within this sequence is metaphorized as river's violence against human engineering, a battle of technology against nature itself in pursuit of the noblest of ends: an increase of the value and productivity of the land. Archival materials will not tell you the extent of the misery and loss of life among Black communities during the 1922 flood—scholars have been forced to turn to a host of cultural objects like music and literature for those insights[50]—but neither does this consideration even appear in congressional testimony. What is the relatively straightforward establishment of a warrant for the laboratory, which is to say, the fairly unremarkable ability to coordinate an argumentative sequence linking the establishment of the laboratory to recent events of note and economic interest, is also a symptomatic appraisal of how disaster could be represented at the national political level. The prevention of disaster was quite simply figured as a matter of reclamation and control of the river and the land. Five years later, the Great Flood caused the federal government to double down on this conception of disaster in the Basin.

THE 1927 FLOOD

The unfathomably disastrous effects of the 1927 flooding of the Mississippi River were the result of a highly anomalous pattern of spring rains across the basin, but the exacerbation of their social effects by political means became quite evident as tales of misery circulated throughout the country. Quite well known on this subject is John Barry's historical monograph detailing the flood and the conditions of its effects, so I will do no more than trace the broad strokes of the even broader political sequence he and others systematically construct.[51]

Since the Spanish and French colonists encountered its mouth in the fifteenth and sixteenth centuries, the Mississippi River has served as a white whale of the modernist desire for control of nature for city planners, engineers, and governments. Famously, LaSalle's introduction to the Mississippi Delta involved not even being able to find its ever-shifting mouth.[52] The river's singular role in the production of commerce and capital—from river transportation to the rich soil it deposits all across the Mississippi Delta—has enticed urban development in its vicinity for three hundred years, and indigenous settlement proximate to it for far longer than that. The tradeoff, of course, is that the climatic factor that produces this soil is seasonal flooding, and the sheer size of the river makes that flooding devastating for permanent settlements.

During the centuries predating European settlement, Native American tribes lived along the Mississippi from the Delta to the upper Midwest. According to Christopher Morris, those in the north were more agricultural than those in the Mississippi Valley because the land was more consistently dry, whereas those in the south lived primarily off the abundance of the land that its wetness provided. While there is some evidence of small agriculture in the south as well, and trade up and down the river existed, life for those in the south was a diverse combination of living according to the patterns of yearly flooding and adaptation to it (by, for instance, constructing giant earthen mounds on which to build). When the Spanish arrived in 1541 with the aims of conquest, they permanently disrupted the economy and political relations between various tribes along the river. Seeing the Valley as unfit for real agricultural settlement because of its wetness, they never settled on a significant scale and were eventually driven out under a barrage of arrows.[53]

Thus, Morris locates the inaugural site of European settlement at the arrival of the French to the Mississippi Delta in 1673, who were at first baffled by the unpredictability of dry land there: "They did not understand it as an amorphous place of land and water. Rather, they saw it as a set dry land that periodically flooded, a misconception that caused no end of troubles as they tried to build on it."[54] As a result, the next forty to fifty years consisted of a series of excursions, forts constructed and then abandoned, and many failed efforts to reconcile their recalcitrant view that they could settle in a French way with the lessons learned from indigenous peoples about how one survives in such an environment. "The French did not come to Louisiana to place themselves at the mercy of the natural environment, or its native populations," he writes.

"They believed themselves superior to both [and] planned to transform the lower valley environment into a dry land suitable for French habitation."[55] And they did, clearing areas for settlement and selecting a site for a port town as close to the Delta as feasibly possible. The land was cleared, and the first buildings of New Orleans were constructed in 1718—only to be completely flooded the next year, prompting founder Jean-Baptiste Le Moyne, Sieur de Bienville to order the first levees to be constructed as a barrier between the river and the town. A major hurricane destroyed two-thirds of the town in 1722. The town flooded again in 1724, this time leaving water for months—and so on.[56]

Control and containment, above all else, has long been the paradigmatic modernist technical fetish when it comes to the Mississippi, and the frequency and severity of hydraulic disasters attests to the degree to which the perpetual failure of such control and containment organized the technical imagination of engineers during the nineteenth and twentieth centuries. There were a number of technological responses to flooding at the time, built or proposed: reservoirs, cutoffs, and spillways were among the options. However, the earliest technology, levees, not only became the most prominent but at one point by mandate became the only federally supported one. Barry writes, "by 1812, levees in Louisiana began just below New Orleans and extended 155 miles north on the east bank of the river and 180 miles on the west bank. By 1858, levees on the two sides of the river totaled well over 1,000 miles."[57] There was a prominent though highly disputed theory that levees, because they restricted the width of the river during flooding, would increase the velocity and force of the river's flow, scouring sediment from the bottom of the river. This would make the river deeper and faster during floods, thus lowering flood stages and keeping the levees from overtopping. Despite all evidence to the contrary, the "levees-only" axiom (whether through mandate, custom, or legislative efficiency) dictated a great deal of flood mitigation strategy for over a century, slowly but inevitably causing levees to be built longer and higher, flood stages to creep up and up, and disasters to threaten far greater consequences with each passing season. As Susan Parrish has detailed,

> Complex and unforeseeable weather patterns across the continent contribute to its uneven flow; storms in the western plains or in the Alleghenies affect water levels in the Lower Valley; the river is perpetually reshaping the land, which, in turn, shapes the river; in all, it is—or was—a regime maintained by its chance-dependent patterns of disturbances. In the course of the nineteenth century, though, major shifts in land use and the built environment designed to make the lands of the watershed more profitable and, in theory, more predictable, nevertheless made the Lower Valley more potentially dangerous to humans.[58]

More profitable and more predictable: the two are inextricable. That the logic of disaster prediction remains coextensive with the logic of racial capitalism is of course one of the basic premises of this book.

The flood of 1927 gave face to the lie that was the levees-only policy. This lie was buttressed by the desire structure of capital:

> [F]or decades the river commission followed a policy of sealing the river off from its natural reservoirs and outlets. This both opened up millions of acres of development, reinforcing the political support for levees-only, and increased the volume of water in the river along with the current. But the stronger current did not seem to dredge out the bottom enough to compensate. Floods that carried less water were rising higher than earlier ones that had carried more. In 1912, for example, a flood devastated the lower Mississippi region. Though carrying far less water than the great flood of 1882, it smashed height records on seventeen of the eighteen river gauges from Cairo[, Illinois] to the Gulf.[59]

Whether a levees-only approach was hydrologically sound (it was not) was far less important than the fact that the policy was politically and economically sound, which is to say no less than that it was designed to support the primarily white-owned and -controlled agricultural sector and the immense profits that were to be had from it in the Delta. In 1927, even this rationale for levees-only dissolved as the river swallowed entire towns and cities, plantations included. Hundreds of thousands, primarily Black and rural-dwelling, were displaced. The Red Cross set up "relief camps" all along the levees, most of which were work camps where Black families were detained and Black men were forced through violence to help the levees contain the water. White landowners "forced as many black flood survivors as possible into these Red Cross relief camps so they could prevent migration and have a readily available labor supply when the waters finally receded."[60] As W.E.B. DuBois argued later that year, in linking the conditions of peonage in those camps to the problem of debt peonage more generally, "Slavery still exists in the Mississippi Valley. . . . The 13th Amendment is not enforced, much less the 14th and 15th."[61] The flooding of the Mississippi, inevitable as it was climatically, had thus become an immense human disaster, and one that reinforces for us the notion that the modernist positioning of disaster is "an administrative problem of political containment."[62] "Levees-only" was a lie because it was hydrologically unsound even for its ostensible purpose, and as a disaster prevention measure it was a lie because its protections depended primarily on the economic ability of individual communities and states to maintain their own sections. Those racially and economically disadvantaged, often placed in more flood-prone areas in the first place, were then forced to live and work upon the levees in servitude to those sections with such racial and economic privilege, just as they had upon the once dry land. That in 1927 the white elite in New Orleans ended up dynamiting a section of the levee downriver to ensure that there would be no breaches in the white portions of the city—sacrificing entire neighborhoods of Black and immigrant populations[63]—is but a horrendous extension of the entire infrastructural logic of disaster prevention and wealth in the region: the proper task of river training and engineering was to preserve the land and assets with the most value to monied whites.

"Levees-only" itself having been dynamited, once more the condition of complexity served an important role in the assessment of what to do about the river. As Andy Horowitz has argued, that dynamite heralded "a regime that enabled politicians and engineers to restructure the earth and how people lived on it in ways that protected and enriched some at the expense of endangering and impoverishing others."[64] The post-flood arguments directly linking the complexity of the Mississippi River with the problematic of land value echoed the logic of the laboratory's advocates of the early 1920s, down to the calculus of disaster. In a debate on the House floor on April 8, 1930, Stephen Wallace Dempsey was pressed to clarify the relevance to the laboratory question of his critique of establishing spillways that "involved the condemnation of a vast acreage of land." I quote his answer at length:

> What was the ultimate and real purpose of the solution of the flood problem? It was that we might make usable the land of the Mississippi Valley. So [with the current approach] we are saving a certain acreage and we are at the same time destroying a certain acreage. When we came to figure out that problem we found that when we offset the value of the land to be destroyed against the value of the land to be saved and then added the cost of saving it, it was a very serious question of whether or not there was any saving; whether or not instead of making a saving we were not making a loss. Then came the purely hydraulic question, such a question as is to be studied under this bill; the question whether we could devise some other way or some other means by which we could instead of destroying land save all the land, perhaps at a greater expense, but instead of saving 1 acre we would save 2 acres; instead of saving an acre and destroying an acre we would have the 2 acres. Now, that is purely a hydraulic question. It is the biggest question facing the United States to-day in cost, in the value of land, and in the extent of the territory involved.[65]

There is, of course, nothing more "purely hydraulic" about the saving of land than about the prevention of the misery along the levees in the Red Cross camps or destroyed communities. The extended reasoning that transforms the social consequences of flooding along the Mississippi into a political and economic calculus of land reclamation asserts the function of hydraulic engineering and delineates the logic of disaster that accompanied the subsequently massive push for a federal flood control project.

From November 1927 to February 1928, the House of Representatives held hearings during which hundreds testified as to the necessity of complete federal control over flood protection along the Mississippi. The argumentative core of this mass of appeals centered on issues of commerce and complexity: the productive potential of the region is jeopardized by its complexity, that is, that decisions in one state affect other states, particularly in times of disaster. The breaching of an inadequate levee in one locale can cause disastrous flooding in another; poor flood control downstream can push floodwaters back up the tributaries into another community; and taken as a whole system, the disjointed spatter of flood control measures across many states reduces both the productivity and the value of all settled land—if not the very existence of some towns and communities. In one such statement, a white Memphis, Tennessee representative complained, "a plantation which last year produced fourteen thousand four hundred and seventy-eight 500-pound bales of cotton this year will produce less than 25 bales of cotton from just a few little patches that we got negroes and tenants to come back in and plant."[66] Moreover, many felt that states and smaller communities had been asked to contribute an outsized amount of land and capital to keep the ever-rising floodwaters at bay:

> The Nation, at its own expense, should construct and maintain levees and other means of flood protection, and should do this without contributions from the localities protected. It is the Nation's stream. The power of the Federal Government over the river has never been questioned. . . . While in the past the local districts have not always recognized the justice of being required to make contributions, yet the record will show that they have been no shirkers.[67]

The appeal to shirking in this statement, made by former Louisiana Senator LeRoy Percy, carries with it a distasteful irony, given that the reduction of the question of "shirking the levees" to an economic metric leaves out any appraisal of the labor of Black bodies on the levees who died protecting Greenville at gunpoint. A Greenville fixture, it was Percy's son William who had given the order to round them up and detain them in the camps, and those who refused were either detained as vagrants or shot.[68] LeRoy himself had previously been involved in a labor scandal involving the illegal debt peonage of Italian immigrants, which his friend Teddy Roosevelt made go away.[69] Nevertheless, white leaders and capitalists across the region made it clear that the time had come for the federal government to take on the responsibility of flood control, at least in the Mississippi Basin, because land was being lost. Not only was the scale of the problem federal by definition, involving many states and many communities. It was also a problem of such complexity that, as one Greenville judge professed, "The plan to be pursued must be determined by engineers. Men who have been able to solve problems of this character, men who have done the work of the United States engineers, can solve this one, and when a sound plan is presented it must be put through for the salvation of the Mississippi Valley and the everlasting honor of our country."[70] Soon thereafter, Congress passed the Flood Control Act of 1928 which, "in the interests of national prosperity, the flow of interstate commerce, and the movement of the United States mails," established a formal flood control program under the discretion of the Chief of Engineers, General Edgar Jadwin.[71]

The inevitable establishment of that project also entailed a reinvigoration of the call for a national hydraulic laboratory, which seemed equally inevitable. However, the near certainty of the passage of the National Hydraulic Laboratory legislation as a clear response to the 1927 flood, as well as the theory of similitude on which it relied, was abruptly called into question by Jadwin in a 1928 hearing. Much like Ockerson had done in 1922, Jadwin laid out a systematic rebuttal of arguments for locating the laboratory in Washington, D.C. that came down essentially to a single metaphor: "You have to keep your feet on the ground and watch the actual things."[72] But unlike Ockerson's, taken as a whole, the testimony performs a functional contradiction as a response to a dilemma that the Army Corps had created for itself: after having opposed the very rationale for laboratory model study for years, it now claimed that it should be granted sole authority over the new laboratory. Whatever institutional motives catalyzed this debate between the Departments of War and Commerce are completely irrelevant to the question at hand; it is manifestly the case that model study as a technology of disaster prediction had become the methodological core of the struggle for this authority. Jadwin's task was therefore to find a way to argue that the laboratory was needed within the geographic and institutional jurisdiction of the MRC and the Corps without ceding too much of the rhetorical terrain that laboratory advocates had sought to claim. The result is largely incoherent:

> Mr. O'CONNOR: So far as the Mississippi and its tributaries are concerned, you do not consent to the doctrine and theory of similitude?
>
> General JADWIN: The doctrine of hydraulic similitude has some points, Mr. O'Connor, and some things have been determined by tests. I showed you where we had done some. But I will say this: That has to be done with reason. You have to keep your feet on the ground and watch the actual things. For example you make the model such a percentage smaller in length and width; then comes the question where you are going to make it the same percentage

> smaller in depth. Then you get your sand in there, are you going to reduce everything but that, shall you take the velocity just as it is in the river or shall you cut that down and call that one-tenth? Those things all come up and they have to be taken within reason. Sometimes you can get a valuable indication; other times if you tie it too closely you will get completely lost.[73]

This is, of course, no answer at all to the question. The task as it was understood by proponents of dimensional analysis was quite obviously to isolate and formalize these sorts of model adjustments, and Jadwin's characterization of the problem unfairly frames them as more or less arbitrary. All of this theoretical abstraction, this working out of the scaling coefficients, he asserts, "it is just prophecies on election . . . it depends a good deal on the equation which you use, and you have to keep watching. You have to keep watching things on the ground, and the nearer you get to it the better off you are."[74] This is ultimately the point—one has to *be there*. Why the fieldwork and the laboratory model are required to be in the same geographical place is left to a loose, enthymematic implication that some coagulation of climatic conditions for each are aligned. What this conjoinment produces additionally, however, is precisely what frames out the future of model studies within the Corps: model study becomes fieldwork once removed, the observation of the empirical conditions of *the model itself*. Afforded a certain generosity of reading, Jadwin's is a fairly reasonable critique of scientific modeling, *per se*. The critique of alluvial river modeling provided by another Corps member during the second part of the Hydraulic Laboratory hearings the following year provides another glimpse: "My conclusion from having seen the operations in the laboratories in Europe is that about 10 per cent of them are mechanical or scientific and about 90 per cent interpretation." This is precisely how Jadwin hoped to transpose the theory of similitude in ways that justified the Corps' work, and as the next section shows, this changed the formalization of modeling itself.

Nevertheless, in spite of the long and concerted efforts of Jadwin and others in the Army Corps, finally, after a decade of formal debates and many years prior of general advocacy, hydraulic laboratory advocates achieved their goal. On May 14, 1930, Congress enacted "An Act Authorizing the establishment of a national hydraulic laboratory in the Bureau of Standards of the Department of Commerce and the construction of a building therefor."[75] True to the Bureau of Standards' primary function, the laboratory was charged with "the determination of fundamental data useful in hydraulic research and engineering," and all projects were required to be formally requested by states or other federal departments. During that very same period, however, Jadwin instructed work to begin on a second, Army Corps-controlled laboratory in Vicksburg, Mississippi that they called the Waterways Experiment Station (WES).[76] Rather than ceding authority to the Department of Commerce, this laboratory would report directly to the Chief of Engineers. In other words, after almost a decade of rhetorical and political labor by Freeman and others to lay out the rationale for the necessity of the laboratory and secure the legislative means for doing so, often against the opposition of the Corps, the War Department quite simply asserted the authority to make one of its own and then did so.

The two facilities would prove to have highly divergent trajectories regarding the hydraulic study of disaster. On February 24, 1931, Freeman complained in a congressional progress report that the staff at the Bureau of Standards had not managed to produce plans for a facility that would meet the needs he had long envisioned for the National Hydraulic Laboratory. Alongside the progress report he attached his own specific designs for a facility located in the Bureau of Standards complex that fit the parameters that Congress had specified as well as one that accomplished his vision—a laboratory capable of researching fundamental principles and that outmatched anything existing in Europe (and not necessarily in that order of priority). The constant and often condescending frustration that characterizes Freeman's discourse more generally is also present in this report. For the $350,000 that Congress had allotted for the facility, he argued, it could be everything that they wanted!

> But, as I have said above, this great opportunity has been sacrificed, apparently through lack of experience of the members of the Bureau staff to whom this problem of design was assigned, and who had no background of practical experience either in large engineering design, or in the special problems of design for controlling or measuring large volumes of water. *The problem of designing this laboratory was intrusted to skillful physicists rather than to experienced and skillful engineers.*[77]

It is possible that Freeman was correct about the faults of the design despite being as self-entitled and controlling as he accused the Bureau of being, for the National Hydraulic Laboratory quickly faded into historical obscurity, while Jadwin's facility became the center of an entire scientific culture of model studies for the next several decades. Whatever the case, the dual irony of this outcome—not only that Freeman's principal appeal throughout the 1920s for the laboratory was theoretical in aim yet apparently so only for engineers, but also that Jadwin's rejoinder held that model studies produced no such knowledge just prior to the WES becoming the premier site of hydraulic model study in the United States—is incidental to the fact that the construction of both facilities at once immediately in the wake of the 1927 flood situates the whole enterprise of physical modeling and the role of models in the establishment of engineering principles as the primary technology of predictive flood control measures from roughly 1930–1970. Indeed, by 1932, a report on flood control along the Lower Mississippi produced by the Corps itself mentions matter-of-factly that "contrary to popular opinion, the establishment of this station [WES] marks no new departure in river engineering. The use of models for the solution of river problems has long been understood."[78]

MISSISSIPPI RIVER BASIN MODEL

Probably to Freeman's characteristic chagrin, the mounting embrace of model study at the WES over the next few decades did not, in fact, result in the systematic revelation of the hydraulic secrets of nature. In 1951, during the very height of model study at the Corps, Chief of the Hydrodynamics Branch at the WES Frederick R. Brown penned a report dedicated to explaining the utility of models "as an aid to the development of design criteria" that began by noting this potential objection: "Hydraulic laboratories are sometimes criticized because they deal too much with specific problems, rather than with research problems of an academic nature which would yield general rules applicable to design procedures."[79] Correlating model and prototype data to achieve the latter, the report implies, is enormously time-consuming and engineers "probably have found the task . . . easy to put off in favor of something else." That data correlation was evidently very boring for these technicians at the very least hints towards the level of investment that the WES had in answering such questions, reflecting and repeating the lack of interest displayed by Jadwin himself. However, without the systematic elaboration of these coefficients, did not model study itself lack a warrant for appearing similitudinous? Was not the entire conceptual edifice that Freeman and his advocates had crafted completely reliant on the necessity of isolating the coefficients, the precise scaling relationships, that such data correlation could produce? Was not model study itself illegitimate without a formalized mathematical ruleset to ensure the accuracy of dimensional scaling? Had not Jadwin argued that the fatal characteristic of modeling was its faith in a deeply flawed conception of similitude and the disavowal of true principles of scientific rigor? Apparently not, as Brown waves away not just the apparent mundanity of data correlation between model and prototype, but also the fact that model technicians might not bother because they "have found inconsistencies for which they could not account."[80] The establishment of coefficients here is posited as a potentially futile and not particularly necessary or interesting venture. Far from being the grounds on which to reject similitude as a theory, these inconsistencies appear rhetorically as merely inconvenient.

If the development of scaling coefficients was no longer necessary as a conceptual legitimization of model study for the purposes of disaster prediction and mitigation, then a different rhetorical construction of similitude was necessary. What I want to suggest ultimately in this chapter is that the conceptual understanding of similitude in physical disaster modeling transformed into a practice entirely and completely dependent on reproducing and re-presenting the *disaster event*. What are posited ostensibly as predictive technologies functioned much more stably and regularly as indexes of collective memory, rhetorically similitudinous reproductions of evental disaster conditions that sanitized, bounded, and ultimately corrected the events that prompted these very technologies. No longer was there much anxiety over coefficients of scaling; changes and distortions of scale could be deemed acceptable through other means.

For instance, while a whole host of WES projects involved constructing scale models of proposed structures to solve very specific technological problems, an initially underutilized approach involved recreating great stretches of actual terrain at model scale. The first such project at the WES was literally carved into the earth in the backyard of the laboratory in 1931, scaled to a degree that would "truly represent the conditions of nature"[81]: a 600-foot model of a stretch of the Illinois River designed to test its backwater limit (how far the river would back up during a flood). There had been some outdoor model studies of landscapes constructed in Europe, but the Illinois River backwater model took exception to those in a telling way: it was far more distorted, and its scale was far more greatly reduced. Fatherree explains that "while European engineers considered a horizontal scale of 1:300 to be large, the Illinois River model's scale was 1:1,200. Its vertical scale of 1:48 produced a 25:1 distortion, a ratio also unused in Europe but necessary to reproduce conditions in so large a prototype as the Illinois River."[82] The sheer size of the rivers in the United States forced the WES engineers to work with radically greater downscaling than in Europe, which also forced them to distort the models equally radical amounts:

> A model of a 100-mile river reach built with a horizontal scale of 1:1,000 would be 528 feet long. For a wide but relatively shallow river like the Mississippi, a 1-mile-wide section of the river would be 5.28 feet wide in the model, an easily constructed and functional unit. However, a 50-foot-deep section of river channel at the same ratio would be a mere 0.05 feet (0.6 inches) deep in the model, and could not simulate flows or other conditions.[83]

"Conditions" is a term left relatively undeconstructed by Fatherree in his explanation of the rationale for these distortions, leaving enthymematic the conditions of veracity for simulation.[84] It is a term often used by the Corps in these contexts and seems to indicate an empirical adherence to the known behavior of rivers. Simulation here is both the methodological challenge and the effect of a representation: for the models to *be* accurate, they needed to *appear* accurate. Consequently, with these scaling ratio increases also came progressively larger projects with even more distorted scaling.

Beginning in 1935, the station purchased land to build what they called the Mississippi River Flood Control Model (Figure 2.2), which used a 1:2,000 horizontal scale and a 1:100 vertical scale.[85] The report begins its theoretical section with what should have been conceived of as a radical departure from the model logic that Freeman championed:

> For studies of flood-control problems, the model must be capable of reproducing accurately the relationship between [flood] stages and discharges. So long as this relationship is maintained, such questions as velocity distribution and path of travel of water particles are not significant. Considerable latitude is permissible in the selection of model scales, and considerable geometrical distortion may be not only permissible but desirable.[86]

Geometrical similitude as a principle is all but thrown out of the rationale completely in favor of the similitude of apparent conditions, and the supplementary logic of coefficients is nowhere to be found. The important relationship here is not the scaling relations of dimensional analysis, but, rather, the relations of input and output and the flood conditions they produce. One might have reminded these engineers of the objections made by their colleague, J. A. Ockerson, who had argued vehemently a decade earlier that "the horizontal and vertical scales should be the same, as variations in the proportions give different results. This would give depth and slopes too small to be satisfactorily measured."[87] However reductive his understanding of theoretical scaling relationships may have been (or feigned to have been), this objection contends directly that if true geometric scaling fails to properly simulate empirical conditions, and if seemingly arbitrary distortions are required to achieve those "accurate" results, then the scaling principle of similitude itself, and thus modeling as a method, is revealed to be of limited value.

These distortions were deemed especially necessary because of the primary warrant for the federalization of flood control: models of individual structures, which had much more modest scaling distortions, had little utility for answering questions about river training for entire systems like the Mississippi Basin. Beyond the direct problems associated with levee failure and breaches, one of the unique aspects of river hydrology is that rainfall, runoff, breaches, reservoirs, or any other element along the system will have effects downstream, sometimes days or weeks later. The Illinois River model, for instance, attempted to reproduce flood stages at various locations simultaneously, hundreds of miles away from one other. It was highly important to know, given particular amounts and locations of input, what the effects of

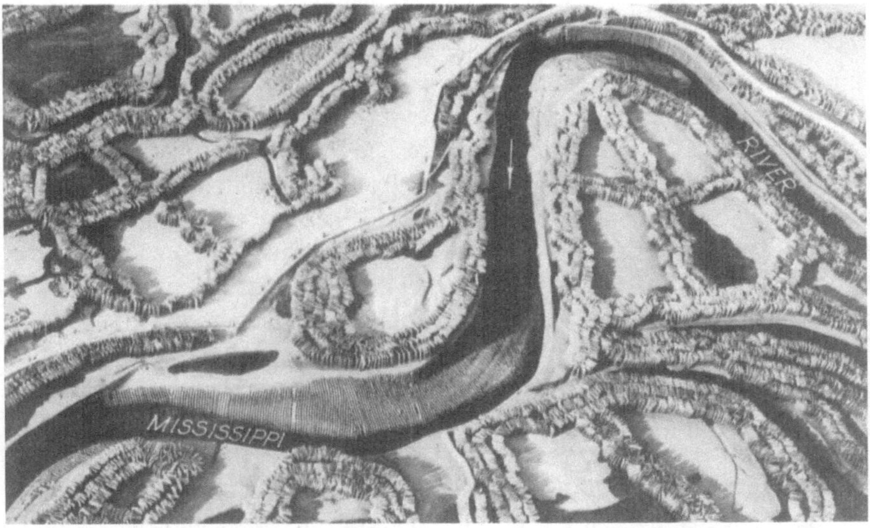

Figure 2.2. Section of model showing wire and stucco roughness, 1942. *Source*: Photo by U.S. Army Corps of Engineers.

certain input and output conditions would be downstream. It was not possible to reproduce an entire system in a physical model without radically scaling it—not only would it be far too large, it would also fail to reproduce accurately the temporal dimensions of the changes in flood stages. That the complexity of entire river systems could not (at that time) possibly be formalized into a mathematical model was taken as given,[88] but even physical modeling, which was posited as a means for either obtaining these calculations or at least approximating them, required the sacrifice of the very logic that had originally legitimized it. Instead, a new logic was pushed to its very limit in the Mississippi River Basin Model (MBM).

The MBM, ostensibly a model system designed to coordinate reservoir operations across the Basin, was specified to the same scaling dimensions as the River Flood Control Model. It was far too large to construct on site, so the War Department purchased a tract of land a short drive away, in Clinton, Mississippi. While the specifications provided by the site already included terrain that roughly approximated the shape of the Mississippi Basin, a relatively large amount of grading had to be completed on the 200-acre site. In 1942, as part of an effort to keep construction costs down and obtain the labor needed to begin the project in a time of labor shortage, the Department established Camp Clinton, a prisoner of war internment camp that eventually housed 1,800 Axis soldiers.[89] Camp Clinton had no other function than to provide labor for the model. The soldiers spent eight hours per day grading, building roads, laying drainage pipes, and clearing trees. According to Michael Robinson, "The food was good and plentiful, and most of the POWs actually gained weight during their confinement. . . . In contrast to supply shortages and rationing within the American civilian community, the interned Axis soldiers enjoyed a wide variety of choice foods including the best grades of beef and butter."[90] This account perhaps too happily accepts the carceral version of comfort built into this labor relationship (an account one suspects the author might have avoided if he were talking about plantation life), but it is nevertheless a description wholly unlike those given of the Red Cross camps in 1927. By October 1945, POWs had excavated almost a million cubic yards of soil and laid miles of piping and roads, sketching out in massive physical form a miniaturized version of the Basin, a "gigantic relief map"[91] that reduced 1.25 million square miles of watershed into 200 acres, and 15,000 miles of stream into 8 miles (Figure 2.3).[92]

Over the next 31 years, the Army Corps would slowly construct the fixed-bed concrete model in accordance with congressional funds and available personnel. Because it was developed sectionally, the completed portions of the model were able to run tests and experiments individually, and any portion could be operated in isolation. This was an important feature and became part of the argument for its eventual completion. In 1952, the Missouri River

Figure 2.3. Mississippi River Basin Model project area looking north, 1 September 1945. *Source*: From the National Archives 2, 77-513-657.

achieved what are still record flood stages across its basin. While the flooding was actively occurring, engineers at the WES ran simulations of the flood conditions in real-time for 16 days, 24 hours per day, advising emergency response teams where to heighten levees, where doing so was futile, and more generally what flood stages could be expected given the reported upstream conditions.[93] These tests were extremely successful, preventing an estimated $65 million worth of damage and quite probably some lives.[94] From about the mid-1950s through the early-1960s, the success of the model during the 1952 flood was routinely deployed as justification for continued funding for the project in appropriations hearings.[95] This recursive relationship between event and simulation, between nature and artifice, prototype and model, was constituted by the pursuit of something like a pragmatic realism, or a practical similitude: the models worked to answer the questions asked, but they did so according to a logic that pinned the authority of similitude between model and nature on the conditions of appearance.

This played out primarily through the process of model verification itself and its relationship to the model's physical appearance. At the most general level, these massive scaled terrain models underwent verification through the circular procedure mentioned earlier: the model is that which accurately explains all the facts, and the true model is to be verified in accordance with the observed facts. Most specifically, though, these models were verified with data obtained from prior disasters. The Illinois River model first ran a series of hypothetical discharge tests (measured in "second feet," or cubic feet of water per second) which, "after a few adjustments," reproduced "the flow of nature to a marked degree of accuracy."[96] Then, having "balanced" the model, the procedure "was repeated for the flood of 1926 (120,000 second feet at Grafton), and for the flood of 1904 (117,000 second feet)."[97] The Mississippi River Control Model report states directly that "the most direct and reliable method of adjusting the various scales of a flood-control model is to first adjust it to reproduce some known prototype occurrence similar to the occurrences to be studied in the model."[98] The verification of the model consisted in part of reproducing a 1937 flood at Vicksburg, and the report also implies bringing to bear the flood stages obtained from 1927.[99] And the MBM lays this logic out in efficient fashion in a 1971 report on the history of its construction:

> A recent flood of considerable magnitude, preferably the maximum flood of record, was selected for each reach of the model to use for verification. The model equivalents of the flows for this flood, from the low stages to the crest, were introduced at the model inflow points and the channel and overbank roughnesses were increased or decreased to make the model stages reproduce those

> of the prototype. With the model adjusted to reproduce past occurrences, it is expected to produce occurrences that may be experienced in the future.[100]

Through these simulations, the models are made systematically to conform to prototype conditions, found "in nature," even alongside the tacit premise that engineering itself has altered what may be understood as nature. Raise the levees here, abandon the levees there, construct setbacks somewhere else—the real measure of disaster lies in the recognition of failure and loss, and these models reimagine "natural disaster" in such a way that neutralizes the political and administrative agency that constitutes its appearance as such. This banal visual reproduction is smoothened or roughened, fonted letters designating each stream and arrows indicating the direction of its flow are embedded in the rivers' beds, engineering structures like bridges and reservoirs are added, folded steel wire screen is installed to simulate the effects of trees and vegetation, all in order to bring the model into direct agreement with nature, both instrumentally and visually. The model itself becomes nature and, in recursive fashion, produces the sense of the natural over and against the rapid and profound technologization of the environment in the so-called prototype. The model memorializes, and in so doing, it deigns to predict.

This is why, contrary to the engineering projects that David Nye identified as evoking the American technological sublime, the predictive physical disaster modeling at hand cannot be understood through the traditional conception of the sublime, as magnificent as both river engineering and the model itself may have been. Slavoj Žižek, in an exposition of Kant, notes that for Kant the terrifying experience of nature's full wrath is what strains the imagination, finds us without adequate capacity to render those experiences into thought. Nature, "in its most chaotic, boundless, terrifying dimension . . . is best qualified to awaken in us the feeling of the Sublime: here, where the aesthetic imagination is strained to its utmost, where all finite determinations dissolve themselves, failure appears at its purest."[101] The imperative to make disaster legible for the subject involves the displacement of that terror and the ascription of its memory to some other object. Kristi Dykema Cheramie, reflecting while standing in the middle of the now-defunct model, however, is "struck by the disconnect that can occur when a model becomes the substitute for the 'real thing,' when the copy, which can never replicate the complexity of its source, becomes the fulcrum around which decisions are made." What we are working with is not the imaginary experience of the sublime, but, rather, symbolic *sublimation* in its somewhat more vulgar sense: the displacement of the traumas of disaster onto an object which serves as its safer proxy, the consolidation of which in our case takes place through the work of rhetorical similitude, a believable likeness or plausible image. Indeed, the model purports to cover the entire Mississippi Basin but, in fact, ends just north of

Baton Rouge—New Orleans was never included. Cheramie correctly notes that "the system was fundamentally incomplete. . . . The Army Corps went on to make 25 years of decisions about flood control here, and modeling the outflow of the Mississippi River never became 'desirable.'"[102] A portent for disaster to be sure.

That likeness in the MBM had been purged of disaster is indicated also by a certain element of popular spectacle: in 1961, the MBM Board realized that "the largest hydraulic model in the world would be of interest to tourists" and so, in 1964, began constructing visitor facilities that included a visitor center, walkways, and a 40-foot observation tower (Figure 2.4). In a 1965 appropriations hearing, the Corps requested funds for, in addition to model studies themselves, "the maintenance of the surrounding area to permit unescorted inspection of the model by the considerable number of visitors expected."[103] By 1971, the WES estimated that roughly 5,000 tourists and technicians visited the MBM per year.[104] It came to be featured on road maps, in the Jackson, Mississippi Chamber of Commerce pamphlets, and in magazines. The spectaclization of the MBM was not the only consequential development that coincided roughly with the completion of the project, however.

The MBM continued to perform tests over the next few years, including a basin-wide testing program to verify the model, but as early as 1970, plans were being set in motion to render it obsolete. In May of that year, the Mississippi Basin Model Board "approved a study to develop computer

Figure 2.4. Postcard depicting visitors touring the Mississippi River Basin Model, n.d. ***Source*: From author's personal collection.**

programs for unsteady-flow computations along reaches of the Mississippi River and its larger tributaries."[105] This began a process whereby the MBM would be retired and replaced almost as soon as it had been completed, thus effectively ending the reign of hydraulic model studies and ushering in another model technology for flood control in the region, "computer application studies." Even more tragic than the short life of the full model was the process by which it would occur: a 1970 congressional appropriations request asserts that "test results previously obtained from hydraulic model tests will be utilized as the primary vehicle for verification of these computer applications."[106] Not only would the MBM be retired, but its final job would be to verify the calculations of the computer programs that would replace it.

This is not really a tragedy, of course, even as the MBM now sits sadly, derelict and overgrown, a handful of miles west of Jackson. The model was an impressive and useful engineering feat of its own in its time, but the condition of its usefulness was always, explicitly, the computational incapacity of modeling fluid dynamics at meaningful scale. With the advent of increasingly advanced measurement devices, the proliferation of measurement as a results of these devices, and the complexification of statistical disaster modeling that resulted from this deluge of new historical storm data, it was only a matter of time before the constraints of physical modeling gave way to new abstractions, and thus new rhetorical configurations of these hypothetical, perfect little storms.

NOTES

1. Foucault, *The Birth of Biopolitics*, 13.
2. For a discussion of the neoliberal manifestation of this economic rationality in the context of disaster capitalism, see Klein, *The Shock Doctrine*, 3–26.
3. John M. Jordan, *Machine-Age Ideology: Social Engineering and American Liberalism, 1911–1939* (Chapel Hill: University of North Carolina Press, 1994), 7.
4. John Jordan, *Machine-Age Ideology*, 3.
5. Mizelle, Jr., *Backwater Blues*, 6.
6. Badiou, *The Concept of Model*, 16.
7. This is not a typo: "prototype," while in colloquial usage tends to refer to a preliminary or experimental model of something, in the world of model engineering signifies the real-world context of application. This usage might seem contradictory, but makes perfect sense according to the metaphysics of the natural outlined in the introductory chapter: the prototype, from the Greek *prōtos* ("first") and *typos* ("impression, mold, pattern"), refers ultimately to the original template that nature provides the model. If an engineer wants to know how forcing a cutoff in a river bend will affect flood stages downstream, they build a *model* according to the original *prototype*—the actual river—and then manipulate the model through experimentation.

8. Mizelle, Backwater Blues, 10.

9. Richard Wright, "Silt," *The New Masses*, August 24, 1937: 20.

10. Lewis Mumford, *Technics and Civilization* (Chicago: University of Chicago Press, 1934/2010), 321.

11. Thomas M. Lessl, *Rhetorical Darwinism: Religion, Evolution, and the Scientific Identity* (Waco, TX: Baylor University Press, 2012), 34.

12. At the same time, Freud cautions against absolute faith in this formulation: "We know that the first step towards attaining intellectual mastery of our environment is to discover generalizations, rules and laws which bring order into chaos. In doing this we simplify the world of phenomena; but we cannot avoid falsifying it, especially if we are dealing with processes of development and change." Sigmund Freud, "Analysis Terminable and Interminable (1937)," in *The Standard Edition of the Complete Psychological Works of Sigmund Freud, Volume XXIII*, trans. James Strachey (London: Hogarth Press, 1964): 228.

13. Herbert D. Vogel, "Origin of the Waterways Experiment Station," *The Military Engineer* no. 352 (1961): 132.

14. Ben H. Fatherree, *The First 75 Years: History of Hydraulics Engineering at the Waterways Experiment Station* (Vicksburg, MS: U.S. Army Corps of Engineers, Engineer Research and Development Center, 2004), 4.

15. Vogel, "Origin of the Waterways Experiment Station," 132.

16. U.S. Congress, Senate, Subcommittee of the Committee on Commerce, *To Establish a National Hydraulic Laboratory*, 67th Cong., 4th sess., 1923, 61–85.

17. U.S. Congress, Senate, Subcommittee of the Committee on Commerce, *National Hydraulic Laboratory*, 68th Cong., 1st sess., 1924, 3–4.

18. U.S. Congress, House of Representatives, Committee on Rivers and Harbors, *National Hydraulic Laboratory*, 70th Cong., 1st sess., 1928, 1–92; U.S. Congress, House of Representatives, Committee on Flood Control, *Abstract of Hearings before the Committee on Flood Control on the Control of the Destructive Flood Waters of the United States*, 70th Cong., 1st sess., 1928, 1–91.

19. Fatherree, *First 75 Years*, 3.

20. For a fascinating account of the troubles that early engineers encountered with this notion, see Layton's examination of the development of war machinery in antiquity: Edwin T. Layton, Jr., "Escape from the Jail of Shape: Dimensionality and Engineering Science," in *Technological Development and Science in the Industrial Age: New Perspectives on the Science-Technology Relationship*, eds. Peter Kroes and Martijn Bakker (Dordrecht, Netherlands: Springer, 1992), 37–38.

21. Layton, "Escaping the Jail of Shape," 39.

22. Badiou, *The Concept of Model*, 16; in the first volume of *Structural Anthropology*, Lévi-Strauss writes, "For though many models may be used as convenient devices to describe and explain the phenomena, it is obvious that the best model will always be that which is *true*, that is, the simplest possible model which, while being derived exclusively from the facts under consideration, also makes it possible to account for all of them." Claude Lévi-Strauss, *Structural Anthropology: v. 1* (London: Penguin, 1993), 281.

23. Ivar Ekeland, *Mathematics and the Unexpected* (Chicago: University of Chicago Press, 1988), 17.

24. Ekeland, *Mathematics and the Unexpected*, 15.

25. Martin Reuss, "The Art of Scientific Precision: River Research in the United States Army Corps of Engineers to 1945," *Technology & Culture* 40, no. 2 (1999): 297.

26. Ekeland, *Mathematics and the Unexpected*, 20–21.

27. Layton, "Escaping the Jail of Shape," 39.

28. Thomas Wright, "Scale Models, Similitude, and Dimensions: Aspects of Mid-Nineteenth-Century Engineering Science," *Annals of Science* 49 (1992): 234.

29. Lord Rayleigh, "The Principle of Similitude," *Nature* 95 (1915): 66.

30. The distinction between static and dynamic in the context of engineering is loose in the first place, as Thomas Wright's essay demonstrates. Dynamical problems abound on apparently static structures; the distinction marks not some essential difference between hydrological and other types of engineering principles, but rather that the rise of hydrological engineering science was roughly coterminous with the evolving understanding of engineering itself in mathematically dynamic, nondimensionalized, but practical, terms; Wright, "Scale Models, Similitude, and Dimensions."

31. Alain Badiou, *Being and Event*, trans. Oliver Feltham (New York: Bloomsbury, 2007), 179–276.

32. Wright, "Scale Models, Similitude, and Dimensions," 252.

33. Margaret Morrison, *Reconstructing Reality: Models, Mathematics, and Simulations* (Oxford: Oxford University Press, 2015), 29–30.

34. "Mississippi River Flood History, 1543–Present," National Weather Service, National Oceanic and Atmospheric Administration, accessed July 19, 2023, https://www.weather.gov/lix/ms_flood_history.

35. John M. Barry, *Rising Tide: The Great Mississippi Flood of 1927 and How it Changed America* (New York: Simon and Schuster, 1998), 156–68.

36. U.S. Congress, Senate, Subcommittee of the Committee on Commerce, *To Establish a National Hydraulic Laboratory*, 67th Cong., 2nd sess., September 3, 1922, 10.

37. U.S. Congress, Senate, *To Establish a National Hydraulic*, 67th Cong., 2nd sess., 4.

38. U.S. Congress, Senate, *To Establish a National Hydraulic*, 67th Cong., 2nd sess., 14.

39. John R. Freeman, "The Need of a National Hydraulic Laboratory for the Solution of River Problems"; in U.S. Congress, Senate, *National Hydraulic Laboratory*, 68th Cong., 1st sess., 3–4.

40. U.S. Congress, Senate, *To Establish a National Hydraulic*, 67th Cong., 2nd sess., 47. Original reads "instance," presumably a typographical error.

41. Morrison's work again is instructive here. Keeping in view the delineation of "abstraction" vs. "idealization" described in the opening chapter is the cleanest way of positing the dilemma from a methodological and theoretical perspective: scale modeling itself involved the idealization of hydraulic phenomena that, in order to be both expedient and accurate, required distortion as a corrective feature. Idealization, rather than complete mathematical abstraction (which, as a reminder, involves the

model itself becoming the primary object of study), allowed for the possibility of similitude. While Freeman's work was accused of being too abstractive, it could be characterized as no less "idealized" than the Corps' field methods.

42. Todd Shallat, *Structures in the Stream: Water, Science, and the Rise of the U.S. Army Corps of Engineers* (Austin: University of Texas Press, 1994), 8.

43. Shallat, *Structures in the Stream*, 4.

44. U.S. Congress, Senate, *To Establish a National Hydraulic*, 67th Cong., 2nd sess., 47.

45. U.S. Congress, Senate, *To Establish a National Hydraulic*, 67th Cong., 2nd sess., 54.

46. U.S. Congress, Senate, *To Establish a National Hydraulic*, 67th Cong., 2nd sess., 4–5.

47. See Huet, *The Culture of Disaster*, 1–4.

48. U.S. Congress, Senate, *To Establish a National Hydraulic*, 67th Cong., 4th sess., 67.

49. U.S. Congress, Senate, *To Establish a National Hydraulic*, 67th Cong., 4th sess., 73.

50. See, for instance, Mizelle, Jr., *Backwater Blues*.

51. Barry, *Rising Tide*; I draw also from Christopher Morris's *The Big Muddy: An Environmental History of the Mississippi and Its Peoples from Hernando de Soto to Hurricane Katrina* (London: Oxford University Press, 2012), 1.

52. Morris, *The Big Muddy*, 27–30.

53. Morris, *The Big Muddy*, 9–23.

54. Morris, *The Big Muddy*, 27.

55. Morris, *The Big Muddy*, 45.

56. Morris, *The Big Muddy*, 70–71.

57. Barry, *Rising Tide*, 40.

58. Susan Scott Parrish, *The Flood Year 1927: A Cultural History* (Princeton: Princeton University Press, 2017), 24.

59. Barry, *Rising Tide*, 157–58.

60. Mizelle, *Backwater Blues*, 33–34.

61. W.E.B. DuBois, "Postscript," *The Crisis* no. 34, November, 1927, 311.

62. Huet, *The Culture of Disaster*, 2–3.

63. See Shallat, *Structures in the Stream*, 128.

64. Andy Horowitz, *Katrina: A History, 1915–2015* (Cambridge: Harvard University Press, 2020), 9.

65. Representative Stephen Wallace Dempsey, speaking on H. R. 8299, on April 9, 1930, 71st Cong., 2nd sess., *Congressional Record*: 6786.

66. U.S. Congress, House of Representatives, *Abstract of Hearings before the Committee on Flood Control*, 70th Cong., 1st sess., 1928, 54.

67. U.S. Congress, House of Representatives, Committee on Rivers and Harbors, *National Hydraulic Laboratory*, 70th Cong., 1st sess., 1928, 6.

68. Parrish, *The Flood Year 1927*, 45.

69. Barry, *Rising Tide*, 13.

70. U.S. Congress, House of Representatives, Committee on Rivers and Harbors, *National Hydraulic Laboratory*, 70th Cong., 1st sess., 1928, 13.

71. Flood Control Act of 1928.

72. U.S. Congress, House of Representatives, *National Hydraulic Laboratory*, 70th Cong., 2nd sess., 78.

73. U.S. Congress, House of Representatives, *National Hydraulic Laboratory*, 70th Cong., 2nd sess., 78–79.

74. U.S. Congress, House of Representatives, *National Hydraulic Laboratory*, 70th Cong., 2nd sess., 79.

75. Act Authorizing the establishment of a national hydraulic laboratory in the Bureau of Standards of the Department of Commerce and the construction of a building therefor, H.R. 8299, 71st Cong. (1930).

76. The original location was meant to be in Memphis, Tennessee. It would seem that the announcement of the relocation of the Mississippi River Commission's headquarters from St. Louis to Vicksburg caused the relocation of the WES as well. It may also be noted that the name—Waterways Experiment Station—quite carefully avoids any of the words "national," "hydraulic," or "laboratory." A detailed account of the administrative politics of the establishment of the WES parallel to the National Hydraulic Laboratory in Washington, D.C. lies outside of this inquiry, and has been performed by others. It will suffice to note, however, that this naming symptomatizes the debate itself, and most of the existing accounts of it are carried out by historians internal to the Army Corps itself. These histories tend to downplay such administrative aggressiveness on Jadwin's part. See, for instance: Fatherree, *The First 75 Years*, 8–11; J. B. Tiffany, *History of the Waterways Experiment Station* (Vicksburg, MS: U.S. Army Corps of Engineers, Engineer Research and Development Center, 1968), I–1–5; Gordon A. Cotton, *A History of the Waterways Experiment Station, 1929–1979* (Vicksburg, MS: U.S. Army Corps of Engineers, Engineer Research and Development Center, 1979), 5–10; Vogel, "Origin of the Waterways Station," 133–34.

77. U.S. Congress, Senate, *National Hydraulic Laboratory Progress Report*, 71st Cong., 3rd sess., 2, emphasis in original.

78. D. O. Elliott, "The Improvement of the Lower Mississippi River for Flood Control and Navigation," U.S. Waterways Experiment Station, Army Corps of Engineers, May 1, 1932, 21.

79. Frederick R. Brown, "Hydraulic Models as an Aid to the Development of Design Criteria," *Waterways Experiment Station Bulletin* 37, Army Corps of Engineers (1951): 1.

80. Brown, "Hydraulic Models as an Aid," 2.

81. Herbert B. Vogel, "Paper Y: Experiment to Determine the Limit of Backwater Influence in the Illinois River," U.S. Waterways Experiment Station, Army Corps of Engineers, February 1931, 2.

82. Fatherree, *The First 75 Years*, 20.

83. Fatherree, *The First 75 Years*, 4.

84. A 1971 recounting of the process of scale selection for the Mississippi Basin Model bears this out as well: "The scales of the MBM were changed from the theoretical scales in order that the lower end of the model, where slopes are small

and velocities are slow, could be adjusted to reproduce prototype occurrences." J. E. Foster, "History and Description of the Mississippi Basin Model," *U.S. Waterways Experiment Station, Army Corps of Engineers*, August 1971, 8.

85. "The Mississippi River Flood-Control Model," *The Experiment Station Hydraulics Bulletin* 5, no. 37 (1942), 2.

86. "The Mississippi River Flood-Control Model," 5.

87. U.S. Congress, *To Establish a National Hydraulic, 1960*, 67th Cong., 2nd sess., 47.

88. A representative of the Army Corps, Lieutenant Colonel Robert J. Giesen, confirmed this in a Public Works Appropriations hearing in 1960, in reference to an upper section of the Ohio River: "Because of the complexity of that particular area of the basin, unless we had the model, there is no mathematical way of doing it; it would be strictly 'guesstimating.'" U.S. Congress, Senate, *Hearings before the Subcommittee of the Committee on Appropriations*, 86th Cong., 1st sess., 1214.

89. Foster, "History and Description," 5.

90. Michael C. Robinson, "Rivers in Miniature: The Mississippi Basin Model," in *Builders and Fighters: U.S. Army Engineers in World War II*, ed. Barry W. Fowle (Fort Belvoir, VA: Office of History, U.S. Army Corps of Engineers, 1992), 284.

91. Foster, "History and Description," vii.

92. Robinson, "Rivers in Miniature," 286.

93. Foster, "History and Description," 27.

94. U.S. Congress, House of Representatives, *Public Works Appropriations, 1960*, 86th Cong., 1st sess., 1214.

95. U.S. Congress, House of Representatives, *Public Works Appropriations, 1957*, 84th Cong., 2nd sess., 1173–75, 2164; U.S. Congress, House of Representatives, *Public Works Appropriations, 1958*, 85th Cong., 1st sess., 968; U.S. Congress, House of Representatives, *Public Works Appropriations, 1965*, 88th Cong., 2nd sess., 1056; U.S. Congress, House of Representatives, *Public Works Appropriations, 1967*, 89th Cong., 2nd sess., 1056.

96. Vogel, "Paper Y," 2.

97. Vogel, "Paper Y," 5.

98. "The Mississippi River Flood-Control Model," 7.

99. "The Mississippi River Flood-Control Model," 8, 11.

100. Foster, "History and Description," 26.

101. Slavoj Žižek, *The Sublime Object of Ideology* (London: Verso, 1989), 230.

102. Kristi Cheramie, "The Scale of Nature: Modeling the Mississippi River," *Places* (March 2011), https://placesjournal.org/article/the-scale-of-nature-modeling-the-mississippi-river/?cn-reloaded=1/.

103. U.S. Congress, House of Representatives, *Public Works Appropriations, 1966*, 89th Cong., 1st sess., 1176.

104. Foster, "History and Description," 29.

105. Billy H. Johnson, "A Mathematical Model for Unsteady-Flow Computations through the Complete Spectrum of Flows on the Lower Ohio River" (Vicksburg, MS: U.S. Army Corps of Engineers, Waterways Experiments Station, 1977), 7.

106. U.S. Congress, House of Representatives, *Public Works Appropriations, 1971*, 91st Cong., 2nd sess., 2275.

Chapter 3

Standard Project Disasters and Rhetorical Transposition

INTRODUCTION

The Hurricane Katrina postmortem—that period during which humanists, scientists, administrators, commentators, and many others asked, "What in the world happened?"—produced a litany of retrospectives, many of which could be accurate all at once. The Bush administration's failure to provide federal relief in a sufficient and timely manner, the inadequacy of flood control design and the legacy of levees-only policies, the geography of racial oppression that routinely relegates Black residents in cities across the country to the most environmentally vulnerable (and often toxified) areas, and many other factors were determined to have contributed to both the immediate and long-term effects of the storm. Ultimately, from an engineering perspective, the failure of the levees at multiple points and the subsequent flooding of 85 percent of the metropolitan area of New Orleans called to account the various federal, regional, and local entities responsible for their design and maintenance. And one of the prevailing lines of attribution took on just this engineer's perspective as media honed their gaze on the simplified and nearly anti-historical query of individual and exceptional infrastructural failure. In this context, "What happened" became more specifically "Why did these levees fail?" Some of the determination of fault lay on what a report produced by an independent levee investigation team coordinated through UC Berkeley categorized as "extrinsic factors" related to failures of organizational and human performance.[1] For instance, it cites the official congressional bipartisan report on Katrina, which noted that both local and federal inspections, while completed on schedule, regularly omitted details such as "brush and even trees . . . growing along the 17th Street and London Avenue canals levels, which is not allowed under the established standards for levee protection."[2] As well, the Berkeley report runs through a long list of organizational failures that indicate the impossibility of maintaining and improving a functional levee system, much of which stemmed from the convoluted manner in which jurisdiction and resources were distributed, lack of leadership, and the imperative to do "more with less."[3] These were all no doubt serious factors. What concerns us in this study, however, is what they term "intrinsic factors," which represent "natural variability and analytical modeling uncertainty," particularly the indictment of the Corps' continued use of a fictional, abstract, and analytically generalized "hurricane" to guide design criteria.[4] The use of this sort of model involved quite a different analytical procedure than with physical modeling.

In 1957, the Waterways Experiment Station produced a physical model and a report on flood-control of the Blackstone River in Woonsocket Falls, RI. The report's primary objective was, for a 2,225-ft reach of the Blackstone River, to provide guidance in improving the channel, replacing the existing dam, and excavation under an existing bridge. The physical model (Figure 3.1[5]) produced for the study was relatively modest compared to the MBM, but was, like many of these later models, nevertheless a similitudinous model of the riverscape itself. What differed fundamentally about the experimental procedure of the Blackstone River model, though, was the introduction of a new concept: the Standard Project Flood. This flood was a pure fiction, a selective, statistically formulated and adjusted composite of, in this case, peak discharge rates likely to result from a severe storm in the region. Although technically a fiction, a flood that had never before taken place, its architecture was determined on the basis of historical data, usually on a regional basis. But the event pegged to the model's experimental purpose—that is, the imagined disaster event, the flood that had never occurred—was no longer precisely a repetition. It was designed as an explicit hypothetical, treated as historical, and then designed against. The storm was invented, and the model tested it.

As the previous chapter demonstrated, disaster was the organizing logic of model study for the half century or so that physical modeling was a principal method of testing design criteria for many hydraulic projects across the country. Within this configuration, disaster was not merely the underside of the project of modernity, which one senses is the way that John Freeman primarily understood it. The derivation and formalization of the laws of nature, for him, was the obvious and universal means by which disasters could be avoided. Rather, as the Corps rose as the premier institution of physical modeling for the purposes of civil works engineering, both the objectives and the warrant for modeling transformed into a far more representational form, strictly speaking. Disaster became the direct means of calibration and veridiction, both the outer limits of design criteria and the internal proof for

Figure 3.1. General view of Blackstone River Flood Control Project at Woonsocket, RI, 1957. *Source*: Photo by U.S. Army Corps of Engineers.

the veracity of physical model study itself. The MBM was only the largest icon of that shift, the case study that took the rationale for physical modeling to the end of the line. The arguments for physical modeling centered on the complexity of these natural relations and the scale of disaster's effects, not only around the Mississippi River but all across the country. Because mathematical modeling was not yet computationally advanced enough to describe these dynamics with any degree of accuracy or efficiency, physical modeling, with all its distortions, contradictions, and challenges, served as a useful approximation. Even as the Army Corps began inventing hypothetical disasters, physical models were for a time still the means by which to test and determine design criteria. It would not take long for the statistical and computational methodologies to begin replacing completely the physical modeling of the early years.

However, the progression of mathematical and computational modeling into new forms and the virtual disappearance of physical modeling within the Corps did not mark an atrophy of the rhetoricity of modeling into more objective or less representational forms, as if the simulations and distortions characteristic of tiny physical models were rhetorical because they were technologically and computationally juvenile. Disaster modeling remained rhetorical both because we need to understand mathematics itself as rhetorical and because, as we shall explore in this chapter, the processes of transposition necessary to stage prediction as a science—to invent a storm and then set it loose over a city or region—however sophisticated its methods, were conceptually independent of those methods. That is to say, these shifts in form and method did not have a necessary or fixed relation to shifts in the concept or operationalization of the science of prediction.

While computational limitations certainly overdetermined the available means of prediction during the middle of the twentieth century, developments in theoretical statistics began also to find a space in the methodologies of Army Corps engineers, particularly those involving probability. The first half of this chapter traces the formalization of these statistical concepts and their integration into disaster science, and the second half theorizes their rhetorical transpositions with similar guiding concerns as the previous chapter: By what scientific rationale were these parameters produced? How did disaster function as an index of rationality in these models? What was the organizing tropology—both scientific and historical—that gave logic to the figuration by which these models attempted to predict and mitigate against disaster? How did political deliberation both impact and document the modes of thought that led to the Standard Projects as an intelligible and authoritative model for producing disaster knowledge? And, in culmination, what are the displacements of public memory at work in these practices? The elimination of extreme disaster events from the dataset and the attenuated retrofitting of data into a bounded temporality of disaster *conditions* (a subtle revision of the "100-year storm" model) obtained scientific indices for the purposes of engineering design criteria, but an exploration of the way these models figure disaster demonstrates as much as the technical upshot of such decisions. The second half of this chapter leverages perhaps the most central relationship within the rhetorical form of this type of modeling—their cartographic visuality and their transposition of disaster across those maps—towards a theorization of rhetorical transposition. Put simply, this type of modeling aligned with a host of post-WWII and early Cold War cartographic rhetorics whose conception of both modernity and disaster—*both*—became structured increasingly through maps. Unlike the unpredictable but geographically static flooding along the Mississippi, hurricanes, much like atom bombs, could be placed, moved, altered, re-centered, transposed, and detonated onto and across maps. And much like atom bombs, a symptomatic core of this rhetorical imaginary was precisely the racial geography of destruction that modeled Ground Zero.

THE FORMALIZATION OF PROBABILITY AND LIKELIHOOD

The question at hand in this section is not an internal critique or refutation of statistical concepts and methods, for, as Hans-Jörg Rheinberger has argued, whereas "experimental traces can be located at the level of scientific object formation, for the historian, experimental systems are the traces at the level of historical object formation."[6] That is, within any given instance of scientific action, one can read discourse traces within the science, symptomatically, to answer questions about the emergence of concepts and truths about a particular scientific object; but to work at the level of a history of science, one must subject the emergence of science's larger systems to the same sort of reading. The rise of the statistical sciences in the early twentieth century is no exception.

There is a practical need within statistical analyses to cultivate data conditions appropriate to the questions one asks. This practical need is the vulnerable core of the most strident objection to its epistemology, that in serving up its own conditions of significance, its own parameters and rules and systems of meaning, in effect its own semiotics, statistical insights are self-authorized. In his landmark text, "On the Mathematical Foundations of Theoretical Statistics," statistician and eugenicist Ronald A. Fisher inaugurates his highly consequential intervention into theoretical statistics with two observations. First, he argues that the nomenclature of statistics unknowingly conflates two quite distinct referent domains: "it is customary to apply the same name, *mean, standard deviation, correlation coefficient*, etc., both to the true value which we should like to know, but can only estimate, and to the particular value at which we happen to arrive by our methods of estimation."[7] That is to say, for Fisher, there is the true *mean* of any given empirical reality, and then there is the *mean* that statistical methods are capable of arriving at which, as a matter of course—insofar as statistical methods function synecdochally by presuming that a part (data set) stands in for the whole (generalized phenomena)—accepts some degree of error. Fisher's complaint is not, however, that statisticians should develop a new terminology for either the real or the estimated concepts, keeping each of these domains nominally distinct. Rather, his insistence on this incommensurability opens a methodological necessity that remains with us today: that error can only be attributed to statistical practice, not to the true value, and so theoretical statistics needs a means by which to calculate the likelihood of this error in any given statistical analysis. He calls this, of course, the p value.

His second observation is, broadly, a description of his perspective on the purpose of statistical methods:

> Briefly, and in its most concrete form, the object of statistical methods is the reduction of data. A quantity of data, which usually by its mere bulk is incapable of entering the mind, is to be replaced by relatively few quantities which shall adequately represent the whole, or which, in other words, shall contain as much as possible, ideally the whole, of the relevant information contained in the original data.[8]

Central to the proper activity of statistics is the systematic reduction of data involving the isolation of which information may be said to be "relevant," and which parts "irrelevant." In this way, one attempts to construct a set of parameters and a mathematical structure that allow for the synecdochal relationship to be established, which is to say, to obtain a sufficiently low *p* value, or likelihood of error. Obviously, the reduction of data to some degree results directly from the questions one asks—whether about wind shear, storm surge levels, or whatever other phenomena one wishes to model. Information that does not sufficiently relate to these aims can be safely culled from the data. This is the absolute simplest way of conceptualizing statistical analysis: the creation of defensible epistemic conditions requires that one reduce data in a defensible manner, and that involves proving a low probability of error. There is more to be interpreted from the argumentative sequence here, though. Why does Fisher's essay move straight from a revolutionary reconceptualization of statistical error directly to an entirely banal observation about the function of statistical method?

The theoretical answer to this question is manifold, and one probably better located in a book about statistical method. The most crucial facet of this answer for the purposes of this book, however, lies somewhere in the midst of both practical and theoretical statistics. Practically, disaster by definition involves an event that transgresses the parameters of the ordinary. The impossible task of disaster prediction, by extension, involves hypothesizing about these as-yet transgressions in order to plan for them. Subsequently, from a statistical perspective, predicting disasters not only involves modeling the frequency and severity of the very events that transgress the statistically probable, but also conceptualizing the parameters of the exceptional within the parameters of the regular—a paradox felt as a definitional contradiction within the very concept of modeling for disaster.

Theoretically, Fisher's critique of the conflation of the true *mean* and the statistical *mean* functioned as a means to pry open a space for a newly formulated conception of error in statistical practice that contained within it a secondary critique—specifically, a critique of the tendency of statisticians to take undue measures to normalize their distribution of data (to force a histogram's symmetry into a standard-looking bell curve), in essence to finesse the data to exclude extremes within any given statistical model. After demonstrating the limitations of a then-common method, he observes:

> This example serves also to illustrate the practical difficulty which observers often find, that a few extreme observations appear to dominate the value of the mean. In these cases the rejection of extreme values is often advocated, and it may often happen that gross errors are thus rejected. As a statistical measure, however, the rejection of observations is too crude to be defended: and unless there are other reasons for rejection than mere divergence from the majority, it would be more philosophical to accept these extreme values, not as gross errors, but as indications that the distribution of errors is not normal.[9]

In this passage, Fisher appears to take aim at one data reduction method in particular, the elimination of values that substantially skew what would otherwise be a more strongly normalized set of data. The offending approach takes the problem backward, Fisher argues, in that its main function seems to be preserving *mean* as the most important parameter, thus organizing the calculation of error around making sure that its distribution is as normal as defensibly possible. Instead, he says, a proper activity of the statistician should be to determine this distribution case by case, not to presume in advance that it will be normal. In sum, he contends that there is no philosophical defense for assuming that the likelihood of an outcome falls into a normal distribution, or an even and symmetrical bell curve. If that is true, then the attendant and consequent requirement of the statistician will also be to calculate the likelihood that the model itself may be incorrect.

At the center of this disagreement, which the reader may at this point be wondering my purpose in spending the last few pages on, is not only what role extreme cases have in the science of prediction. Such a question has clear and obvious implications for the question of disaster prediction, policy, and infrastructure. The center is also, at root, a terminological and conceptual balancing of *probability* to *likelihood* and the attendant logics therein. *Probability*, essentially deductive, describes what Fisher calls the "frequency ratio" of something happening based on established parameters, and its object is the outcome of the phenomena in question. The classic example is a coin toss: we know that there is a 50 percent chance of a coin landing on heads in one toss, a 25 percent chance it will land on heads two tosses in a row, and so on. That is why one might cull extreme values in a given data set (for instance, a data set that contains a completely unlikely sequence of 100 heads in a row): it messes with established laws of probability, thus also skewing the possibility that the prediction is wrong. But this deductive movement relies completely on the conflation with which Fisher begins his essay—namely that, in any actually existing context, the presumed probability is coextensive with the probability that statistical methods are able to calculate. The internal justification is nothing more complicated than the assumption that the coin *should be* fair. When one starts with the assumption that the probability of a coin toss is, in fact, 50/50, one will necessarily end up with a normal distribution of error. Culling an observation of 100 heads in a row becomes a matter of bringing the dataset into better conformity with the model.

For Fisher, when determining *likelihood* from a statistical perspective, the line of inquiry moves in the opposite direction, which can determine one of two things. First, the observed data in this example is used inductively to answer the question, "*is* the coin fair?" Given the previous example, 100 heads tosses in a row is so unlikely given the established 50/50 model that we are given to assume the coin itself may be rigged or defective—not that it's impossible, but that it's so improbable that it seems more likely that the tails side must be heavier, or that the coin was secretly swapped for one with two heads sides. For situations like a coin toss—the paradigmatic coin toss, a standard colloquial matter for something that could go either way—the model assumes a 50/50 probability in advance. When we toss a coin, the operative desire structure is one that *really does want* the coin to be fair, so we may determine likelihood to ensure that this is the case, that the game is not rigged. However, and this is the second possible determination, many times we are interested in shoring up trust in the model itself—to what degree is the model correct? Whereas the coin toss is posited for all time to represent a normal distribution resulting from a 50/50 probability, and this should be borne out with a greater and greater number of coin flips, there is no such assumptive luxury involved in predicting disaster conditions, or, really, most other kinds of events. Instead, this type of statistical practice reduces a massive but finite population of data into what has been determined to be a representative sample—I have called this synecdochal. But then it explodes this representative sample out into a hypothetical infinite population, both hypothetical and infinite because within the formal dimension of prediction there is nothing to observe.[10] So, as opposed to answering whether the coin is fair, the purpose is to use the existent data to ensure that the *model itself* is true. Now one sees the necessity of including extreme values, for they allow the researcher to arrive at a more accurate answer to the base question: what if the model is wrong? Or, perhaps more precise than "what if," *what is the likelihood* that the spectral reduction of data in any given statistical model is erroneous in the face of actually-occurring future disasters?[11]

On the ground, as it were, this is more a practical problem than a conceptual one. The transgression posited by disaster represents a violation of normal conditions, and the variables involved in that violation vary from place to place, from disaster type to disaster type, and from one acceptable level of destruction to another. What factor should be used as a threshold, and what point must be exceeded to recategorize an event as extreme? Furthermore, as with mitigation efforts along the Mississippi, what are engineers willing to accept as upper limit of protection in any given context? These questions appear to demand an answer in the form of a physical variable: wave height, or storm surge levels, or water stages, or rainfall rate, or some other indicator used as the central determinant of intensity or extremity. In the case of hurricanes, as we shall see, it was a generalized "storm intensity" indexed by central pressure indices.[12] But this kind of threshold is only one factor, because the language of probability for disasters involved not just measures; it necessarily also involved time. In drafting these thresholds of acceptable protection, one needed to decide, for instance, how high to build the levee, and to accomplish this one needed an idea of how often storms of what extremity would occur. Time and time again, Congress and the Army Corps confirmed the impracticability of designing and constructing protection against any and all disaster conditions. It was an economic calculation, certainly (the expense of the structure vs. the expected expense of destruction), but that calculation was one of temporal probability: what are the chances that a storm of a particular intensity will exceed those conditions within a given timeframe? Was it really worth building an infallible solution where costs may be prohibitive?

Thus, the operationalization of the *return period* (or *recurrence interval*), colloquialized normally as something like a "100-year flood," became a central calculation in crafting simulated storms. As Resio et al. remind the reader, this is very much a calculation of probability: "it is very important to recognize that a *100-year event* is not constrained to occur only once every 100 years. . . . Over 30 years the probability of encountering a 100-year exceedance would be 0.2603 [26.03%]; and over 50 years the probability of exceedance would be 0.3950 [39.50%]."[13] In concrete terms, that means that a structure built to mitigate against a 100-year event stands a not insignificant chance of being tested against those conditions within its first few decades, even if it was built directly after a 100-year event had occurred. For the same reasons, the decision to design against a 100-year event rather than a statistically much less probable and more extreme 200-year event still faces a 26 percent chance of nevertheless encountering a 200-year event within the first 60 years of a structure's lifetime. Depending on the allowance one is willing to accept, the return interval necessary for reliable protection rises at an almost alarming rate. In fact, as Ven Te Chow's highly influential volume *Handbook of Applied Hydrology* detailed in 1964, "If, for example, the design engineer wants to be approximately 90 per cent certain that a 10-year or greater rainfall does not occur in the next 10 years, he [*sic*] should design for the 100-year rainfall."[14] In other words, as we will see in a later chapter, mitigation was positioned as merely a question of "when" rather than "if," and relatedly of "how severe."

All of those concepts, constraints, and imperatives now detailed, we may turn to the simulations of the middle of the twentieth century. Unlike the perfect little storms of the physical modeling era, which were by and large sanitized reproductions of historical storms, the simulations emergent during this time period were purely hypothetical, consolidated fabrications. Instead of simulating structures and landscapes, and then using historical storm measures as inputs for model verification, engineers simulated the yet-to-occur storms themselves on the basis of those same historical storms. Conceptually, these disasters may be considered hypothetical in two related senses. First, they were hypothetical in a statistical sense, which is to say they were modeled for an infinite population of possible outcomes, an infinite number of points along the probability scale. They were also hypothetical in the more colloquial and rhetorical sense, indicating the spectral, the pre-figural, the disasters that statistics dreams up and wonders about. They were posited to exist in the abstract but used as a framework for the potential-actual; they were made manifest through rhetorical approximation without any presumption that such conditions would ever actually manifest down to the detail. Sometimes, in fact, that would have been impossible.

STANDARD PROJECT STORMS

Beginning in the early 1950s, the Corps embarked on this new approach to studying and determining design criteria to withstand catastrophic climate events. Anchored by the ideals of generalizability and standardization, they proceeded to invent three transposable disaster models: A Standard Project Storm (SPS), a Standard Project Flood (SPF), and a Standard Project Hurricane (SPH). The names and definitions of each of these models indicates the structure that unites them. A 1952 report published by the Corps defines a Standard Project Storm as an estimation that:

> for a particular drainage area and season of year in which snow-melt is not a major consideration should represent the most severe flood-producing rainfall depth-area-duration relationship and isohyetal pattern of any storm that is considered reasonably characteristic of the region in which the drainage basin is located, giving consideration to the runoff characteristics and existence of water regulation structures in the basin.

The same report gives the following definition of the Standard Project Flood: "In general terms, the standard project flood may be defined as a hydrograph representing runoff from the standard project storm (and/or snow melt)."[15] The two operate hand-in-hand: the SPS is conceived as the most severe but also "reasonable" storm one can calculate for a particular region and time, and the SPF describes broadly the flood conditions that result from that storm. Other factors like flood infrastructure and terrain contribute to the parameters of the model, but the aim of these calculations is the production of a make-believe storm and the flood it conjures, which then provide broad parameters for predicting future storms and floods. Seven years later, the Corps set out to do the same with the Standard Project Hurricane: "Like the Standard Project Storm precipitation, the SPH index is based on enveloping the records of meteorological events with elimination of a few extreme events. . . . The SPH index parameters . . . are consistent within the limits imposed by regional variation of climatological features."[16] In less technical (and less precise) terms, extant historical data (1900–1956)—both measured and approximated indirectly—for hurricanes in any given zone were plotted statistically in such a way that the Corps could establish a "worst-case" scenario that was nevertheless guardrailed by statistical "reasonability." These fantasmatic "hurricanes" were then used to inform design criteria, in any given zone, on the basis of a 100-year return period of a storm based on its central pressure index, which is the measurement used to stand in for intensity.[17]

A central precondition for the emergence of the SPS and the SPF as scientific rhetorical forms can be found in 1928—even before ground had been broken on the Waterways Experiment Station—in a flood control plan that General Jadwin submitted to Congress. In that report, Jadwin, driving home the end of the levees-only era, acknowledged that the task required that the nation "not try to restrict the Mississippi River too much in extreme floods. The river will break any plan which does this." Moreover, "the levees that have been constructed are not sufficiently high for floods that are now predicted."[18] A good portion of the plan involved integrating reservoirs, spillways, and other structures that can temporarily increase discharge capacity without physically restricting the river past maximally defined flood stages. Strengthen the levees and introduce additional measures to reduce reliance on them. How to define "extreme," though? On what basis might the Corps design against a flood as yet unrealized, unspecified? The rudimentary answer to this question lies within the vote of confidence Jadwin gives his Corps' own plan:

> Should Divine Providence ever send a flood of the maximum predicted by meteorological and flood experts as a remote possibility, the floodways provided in the plan are still normally adequate for its passage without having its predicted heights exceed those of the strengthened levees. The plan is designed to be both simple and comprehensive, flexible and adequate to prevent a calamity such as that of 1927 from happening again as the result of any flood past or predicted.[19]

Prediction, perhaps reasonably, becomes the scientific procedure upon which these determinations must necessarily rest. One cannot arrive at standards for protecting against disaster unless one anticipates the conditions that the disaster will produce. What is omitted completely from the Jadwin plan are the precise methods for determining such concepts as "maximum," "remote possibility," or even "calamity." Despite indication otherwise by some Army Corps historical accounts (which sometimes draws a straight line between the Jadwin plan and the SPH), by 1928, there was no core technical architecture yet articulated within the Corps that designated which parameters of disaster should be specified, how one should calculate those parameters, and the extremity of their hypothetical weather conditions.

It was not until the 1952 report establishing the conditions and parameters for modeling the SPS and SPF that the Corps attempted their definition in these precise terms, and two items therein warrant our attention. First, the report makes a crucial distinction between the "Standard Project Flood" and the "Maximum Probable (or 'Maximum Possible)' [*sic*] Flood."[20] The SPF, it will be remembered, represents the most severe flood conditions "that are considered *reasonably characteristic* of the geographical region involved, excluding extremely rare combinations" (my emphasis). The Maximum Probable (or Possible) Flood, on the other hand, represents the most severe flood conditions "that are *reasonably possible* in the region" (again my emphasis). As far as definitions go, this distinction leaves a lot wanting. Reasonably characteristic vs. reasonably possible? At issue is both the definitional split itself and the figure of reasonability.[21] In practical terms, the Maximum Probable Flood describes a hypothetical upper-limit event, a perfect storm, as it were, whereas the Standard Project Flood describes a fairly (though not maximally) disastrous flood, a "practicable expression of the degree of protection that should be sought as a general rule in the design of flood control works."[22] The distinction is a direct acknowledgement that the construction of mitigation infrastructure perceived the extremity of disaster at two levels: one that was to be stopped, and another that was not. This exposes the figure of reasonability, at least when it came down to the protection of life and land, as a quality not of the storm itself, but rather of the calculus of allowable destruction—offloaded onto a temporal measurement of return periods—determining whether disasters would be allowed to take their course.

In practice, the challenge of these calculations was eventually to determine both what mitigation infrastructures and procedures to construct or implement for any given locale or region and, most importantly for this point, just how severe a weather event those technologies should be built to withstand. To do this, one must determine the characteristics of a semi-worst-case weather event that could occur in a given region—the Standard Project Storm assigned to that region (semi-worst-case because, again, this was a threshold that ignored the most extreme events). As with the Mississippi River models examined in the last chapter, the means for determining the parameters of each category was by strategically collating historical data into conjectural data. The 1952 report provides an extended example of an estimation made for an 11,700 square mile area of the Cumberland River basin north of Nashville, TN. Using Army Corps historical data, "approximately 30 storms of record within a few hundred miles of the Cumberland basin were tentatively listed for transposition" based on rainfall totals exceeding 6 inches over 5,000 square miles.[23] From this list, researchers determined that winter storms created more severe flood conditions, so the list was culled to 19 storms, which were then "transposed"—which is to say, moved cartographically over to the center point of the region in question—twice, once to the lower and once to the upper parts of the Cumberland basin. Then, using methods and metrics unspecified, researchers isolated the seven "most important" storms of these 19 for further study. Finally, having made statistical adjustments to each storm to account for differences in average rainfall, basin shape, and other relevant flood conditions, it was determined that a single storm—the storm of March 23–27, 1913—would serve as the Standard Project Storm for the area in question. It is notable, if not entirely symptomatic, that of all 19 options, the chosen storm was also known by another name, the Great Flood of 1913, which killed over 700 people in the Midwest, left millions more homeless, and was second in social and economic consequence at that time only to the famous Johnstown Flood of 1889.

The statistical reduction of data, from an initial threshold of 30 storms to just one, is the first half of the methodological sequence here and, as we have said, is perceivable through the lens of synecdoche. *The Handbook of Applied Hydrology* explained in 1964 that "in ordinary cases" (which we might assume means "as a general rule") the use of a single storm of record, with adjustments, is likely the method one should use. There were less ordinary conditions that might warrant the construction of a Frankensteinian assemblage of several storms into one hypothetical pattern, but the general suggestion was that researchers use the transposition of empirical data from a single storm as representative. The authors go so far as to suggest that this does not just approximate the characteristics of a given region—the "*experience* of a basin is extended through transposition."[24] It is as if the historical data itself has been conjured, as if by moving a storm from one place to another, one can craft a memory for a drainage basin that has had no such recorded experience itself.

It is no surprise, then, that the second half of this sequence might be understood as a kind of retrofitting of synecdoche—the use of scientific methodology and the alteration of the data space as means of proving the synecdoche.

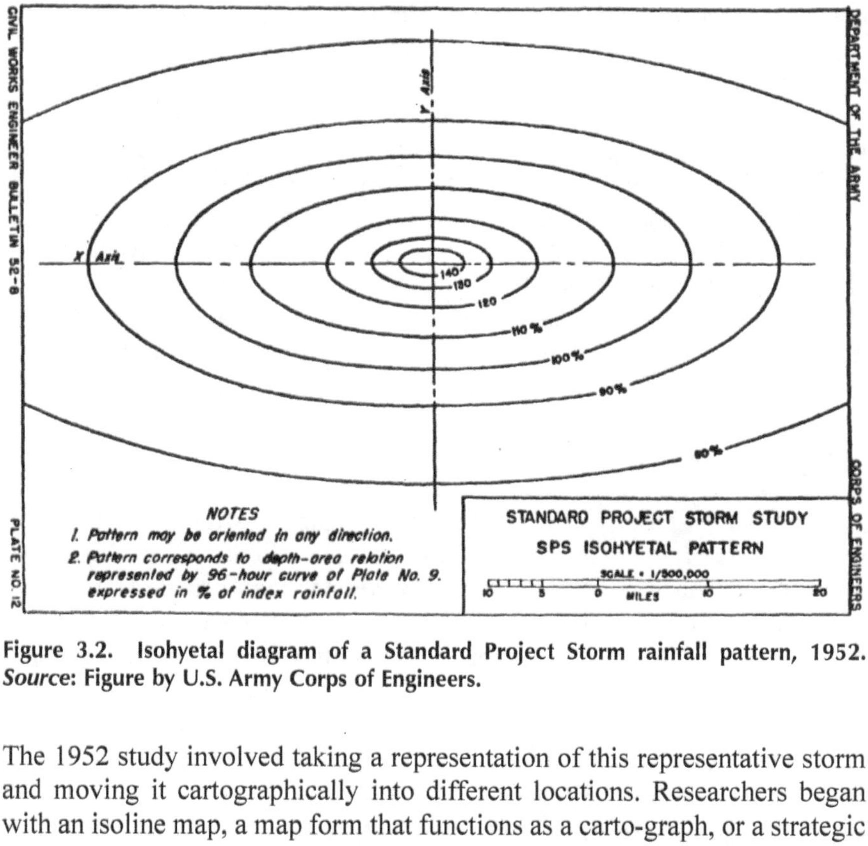

Figure 3.2. Isohyetal diagram of a Standard Project Storm rainfall pattern, 1952. *Source*: Figure by U.S. Army Corps of Engineers.

The 1952 study involved taking a representation of this representative storm and moving it cartographically into different locations. Researchers began with an isoline map, a map form that functions as a carto-graph, or a strategic representation of space that superimposes connected lines, in this case concentric, to describe something in spatial terms (Figure 3.2). In everyday contexts, people most frequently encounter isoline maps, which often resemble rings in the cross section of a tree, when looking at elevation on topographic maps. In the case of the 1952 report, the isohyetal rings denote precipitation values, or how much rain each specific location on the map received in the given storm period. Researchers then superimposed this isohyetal map of the 1913 storm on top of the new area to be studied (see Figure 3.3)[25]—except one cannot justifiably move a storm hundreds of miles without alteration of the data space, both statistically and cartographically. Just like the principle of similitude and the scaling woes of the engineers in the nascent industrial age, extra calculations were made to account for differences in average rainfall, basin shape, the amount of rainfall the soil would accept, and the existing drainage capacity of the basin.[26] These adjustments were referred to as "transposition factors" and functioned to ensure that a standard project storm conformed to the basic regional characteristics of the region in question.

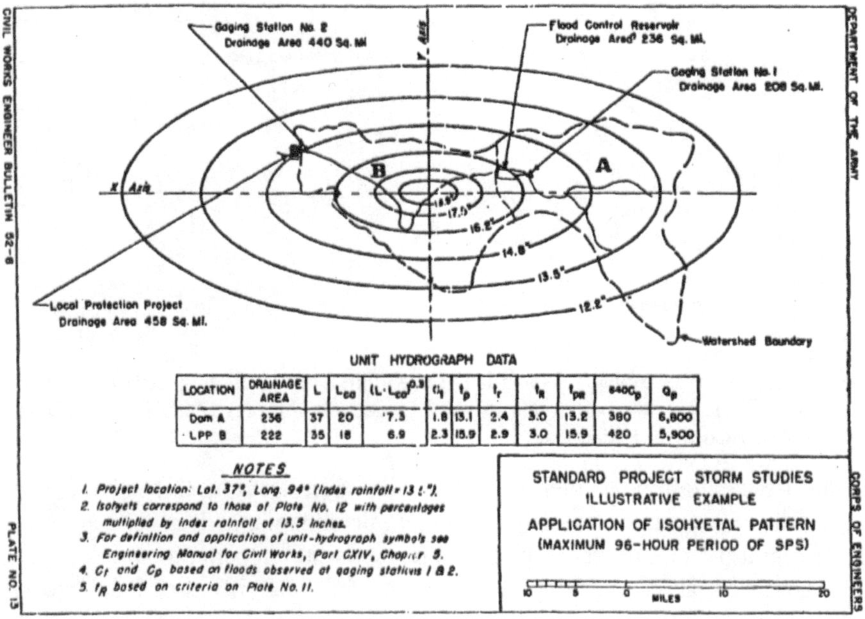

Figure 3.3. Isohyetal diagram of a Standard Project Storm rainfall pattern positioned over an existing watershed, 1952. *Source*: Figure by U.S. Army Corps of Engineers.

The elimination of extreme events from the data set, called enveloping, helped make smoother curves by evening out the historical pattern (see Figure 3.4). This procedure, statistically speaking, involved three steps: first, the calculation of probable maximum values of storm intensity for the region in question through a statistical analysis of historical storms;[27] then, the selection of a single historical storm (or, in rare cases, a stitchwork of several storms) on the basis of these values; and finally, an adjustment of the historical data (which is to say, a re-sizing, a re-intensifying of the storm) to bring it into tangent with the initial probable maximum values.[28] The graphical curves—the envelope—created by plotting this data space mapped an arterial zone of probability with an upper and lower threshold, and the transposition procedure involved, at the center of its guiding logic, refitting the past to be perfectly overlaid on the top of those thresholds. Once the old storm fit snugly into its new space, all the other relevant calculations necessary to determine engineering design criteria followed deductively.

A few rhetorical and epistemological problems arise. In the first place is the elimination of events with extreme parameters, about which we have already covered the philosophical objections levied by Fisher. More centrally, though, what epistemological principles justify "relocating" a massive,

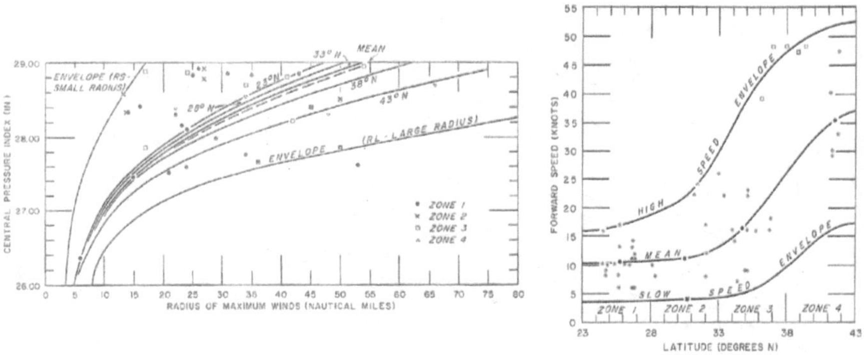

Figure 3.4. Two examples of statistical enveloping, one modeling central pressure index and wind, the other modeling forward speed and latitude, 1959. *Source*: Figures by U.S. Army Corps of Engineers.

once-in-a-generation weather event 330 miles from Dayton, OH to Nashville, TN? This concern is necessarily methodological insofar as it is epistemological. From a rhetorical perspective, this transposition occurs not simply in the data space of tables and graphs, in the warping and fitting of data into its graphic visualizations, but also across cartographic space: it is the movement of a particular representation of weather and risk across miles and borders, and just as consequentially is the pause and detonation of those storms onto a predetermined radius of effect. It is a Cold War tropology in full force.

Transposition

When Hans Jörg Rheinberger argues that we cannot have the whole of an experiment, he gestures implicitly towards the trope of synecdoche that we've provisionally assigned to the reduction of data that is the selection of standard project disaster objects (storms, hurricanes, floods). The essay out of which this claim comes, however, is much more centrally organized around a different rhetorical concept, transposition, that operates for us at two distinct levels.

For Rheinberger, transposition signifies first the movement between what he understands to be the experimental *trace* and the representational forms that *data* takes on in the course of the consolidation of experimental knowledge. "Transposition basically means that things are taken out of a particular context of use," he argues, "and brought into a constellation where we can marvel at and do things with them."[29] This is a movement from an epistemic space—understood through the Derridean concept of the trace, whose "origin" is constituted retroactively by the process of tracing itself—to what Rheinberger understands to be a representational space. In short, experimental data (or in our case historical data) are given representational form: the technical activity involved in data generation is transposed into more fixed, manipulable, but bounded forms. Tables, charts, graphs, isohyetal maps, physical scale models—all of these are transpositions into and within a data space, which then takes on the appearance of a neutral medium attempting to report disinterestedly on the data points in question. It takes the physical and observational data of experimentation and observation (already itself overdetermined by the technologies of experimentation and observation) and reproduces them in forms that claim, to paraphrase Rheinberger, to contain the core *information* of the object in question. It is not necessary at this point to argue through why data spaces can never be truly disinterested or neutral spaces of data reporting. Most important to Rheinberger's use of the Derridean trace is one of the most basic of all Derridean insights: that the appearance of such neutrality, such transparency of conveyance, is produced by the very activity of its writing. To that end, one senses that the word Rheinberger grasps at might be "signification." But that is not all.

The second-order activity of transposition, according to Rheinberger, happens *within* these data spaces in the process of modeling. By virtue of their very purpose, he argues, they "operate in and through a medium that presupposes an ontic cut with respect to the target phenomenon, the epistemic thing in question."[30] That is to say, whereas experimental and observational scientific practice normally does its utmost to force the technics of experimentation and observation to "disappear," as it were, from the outcomes of the activity, in modeling it is acknowledged outright that these outcomes do not contain the information of the epistemic thing. Instead, this thing is recreated in another context; it is a knowing fabrication. It is a fabrication, however, that relies entirely on inversion: following Lévi-Strauss, Rheinberger notes that conceptualizing a model involves a conceptualization of the whole before the individual parts are known, or the creation of a whole in the service of the determination of its parts. And even if this perception of the whole is phantasmatic (as it must be, in models that operate hypothetically), it is their proper function to transform clusters of data into something that appears as a self-contained whole, a reduction of data, a simplification that fixes the relations of its components to one another such that one part can be manipulated to watch those relations respond and transform. As a result, he argues, "we cannot have the whole in an experiment. The permissibility of such simplification is up for permanent negotiation."[31] Science rips data from its context yet requires that it remain intact unto itself. Together, this two-part transposition is first the transmogrification from experimental trace to data (representation or, more strongly, signification), and then the outfitting of that data space into a manipulable but entirely and intentionally out-of-context whole.[32]

Rheinberger's two-part schema is useful for precisely the same reasons that invite reservation. On the one hand, using transposition as a master concept for theorizing the production of scientific knowledge helps the critic position, as it were, the various historical elements of scientific methodology and the construction of its epistemologies; in effect, it allows the critic to trace the manipulations and triangulations required for knowledge to appear as such and simplifies that critique under a central rubric. On the other hand, however, there is a certain dulling of the concept at work, which I would attribute primarily to Rheinberger having peeled the dynamic of the trace/non-origin away from the theorization of signification and rhetoric. Indeed, he notes up front, the use of Derrida's trace make it "possible to determine more specifically what goes on in the material transformation processes of *scientific experimentation* in terms of epistemic procedures, without always having to deal with the heavy load of critique accumulated on the notion of representation."[33] This is at one level a prudent choice: to trace the historical progression of experimental science and its procedures requires pulling critique out to the level of object formation and the recursivity involved in the notion of progression itself. And his careful examinations of the transposition from trace to data in, say, a viral DNA sequencing procedure that renders a graphematic, alphabetical string as the end result of an experimental method do give the sense that rhetorical form and representation are consequential within his theorization, that transforming measurements from a historic storm into a graph instead of an isohyetal map, or vice versa, is a difference that matters. So, my reservation not that "transposition" does too much work in Rheinberger's schema—how much latitude have I given the term "model" in this book?—rather, it's that any reconfiguration of or resignification within the data space is understood in the end as transposition, but once this conceptual move has been made, there are few means to explain these differences that matter.[34]

One central consequence of the desire to fit and smooth models in the process of statistical transposition, for instance, is that fixing parameters into a hypothetical storm does not always accurately predict the most consequential factors in hurricane events. As Resio et al. have argued, in addition to the vagueness in defining SPH and PMH as concepts, reliance on "*sustained wind speed* as a fundamental parameter . . . implies that the highest sustained wind speed will always produce the highest storm surge." In fact, they note, while Camille dwarfed Katrina in this category, the storm surge levels—the primary factor when it came to infrastructural failure—were much higher during Katrina.[35] Pegging the parameters of a storm to "intensity" (measured by central pressure) or surge levels to sustained wind speed gave engineers a formalized basis for design criteria when, in actuality, variations in these relationships is the norm. From a statistical perspective, particularly keeping in mind the computational limits of that time period,[36] it was defensible as an approximation. As computational capacity, sample size, and theoretical advancement have progressed, these early decisions become easier to critique. The rigid formalism of these calculations led design criteria astray, failing to account for storm conditions occurring outside of the pre-set parameters of the standard project models.

But from a rhetorical perspective, this particular formalization of storms—which most certainly did, on a purely practical level, compensate for a lack of computational ability—also resonated deeply within another, not unrelated field of disaster modeling. That is, the impulse towards the transposition of the source data into a particular cartographic form changed fundamentally the exercise of prediction by reconceptualizing storm disasters as simultaneously both mobile and fixed. And why not? If the methods (and thus rhetorical forms) for determining infrastructural design criteria for storm mitigation were considered to be sound, then it's reasonable to posit that these methods could be used to design against other disaster events—like atomic bombs. If it is not so far-fetched to postulate that the rhetorical forms of disaster modeling—both the statistical formalizations and the cartographic ones—could resonate across seemingly disparate disaster scenarios, then it is likewise easy to assume that these resonances are symptomatic of a greater epistemic affinity about the nature of disaster. Disasters were, almost suddenly in the post-WWII era, hostile deliveries of variable intensities whose radii of destruction gathered towards an increasingly violent Ground Zero.

CARTOGRAPHIC TRANSPOSITION

It is by now well established in the historical literature that the emergence of civil defense culture in the United States brought war to the homefront in myriad ways: through industry, the patriotism of commerce, the "dedensification" of cities alongside white dispersal to the suburbs, and even the construction (or at least the civic imperative towards construction[37]) of residential fallout shelters. The specter of nuclear warfare was both familiar and entirely novel, and it might be argued that the distinction between natural and human-caused disasters within the U.S. catastrophic imagination began a slow process of transformation with the advent of the atom bomb. In the 1940s and 1950s, this was not primarily a matter of "mutually assured destruction" (that development would come in the 1960s as the bombs grew in power and number)—rather, the disaster models of these early nuclear decades were highly organized through cartographic visual means, which is to say that the atomic bomb remained firmly a matter of place and location: Ground Zero and its radius. Ground Zero became the central metaphor for the center of destruction, and was used liberally in the 1940s and beyond to refer to the hypocenter of an atomic blast, or the point on the ground directly below the epicenter, "Air Zero." In fact, Paul Boyer's well-known historical exploration of the nuclear age, *By The Bomb's Early Light* (1985), argues quite directly that with few modern exceptions, "every theme and image by which we express our nuclear fear today has its counterpart in the immediate post-Hiroshima period." It's an extraordinary claim to historical continuity, finessed perhaps a bit too forcefully by the ambiguity of the word "counterpart." But it remains an intelligible claim, insofar as this perception relied heavily on recognizable rhetorical forms imported from longstanding discourses and adapted for context.

The utter destructiveness of the bomb, for example, required a frame for imagining its destruction different from that of traditional warfare, and experience with natural disasters lent that form. As Fred Charles Iklé noted in 1958, while conceptualizing the social ramifications of urban atomic detonations:

> While test explosions give nuclear physicists an opportunity to measure physical effects directly and thus to confirm or revise their theoretical calculations, such experiments are not available for studying social effects. Social predictions must, therefore, be based exclusively on an analysis of the fundamental relationship of physical destruction to social phenomena. The intricacies of this relationship must be derived largely from the study of actual disasters; they cannot be deduced solely from general principles of sociology, economics, or other social sciences.[38]

In other words, without precedent for predicting what would happen in the U.S. in the period immediately following an atomic blast, Iklé reached for a reasonable analogue. And as David Monteyne has documented, the nevertheless "incomprehensible effects of nuclear warfare" forced its imagination to be "rendered meaningful through comparison with the familiar, such as the streets and structures people knew from their everyday lives. In particular, natural disasters provided a ready-made set of images, fears, and responses that could be deployed to explain the nation's new, complicated, and possibly dangerous role on the global stage."[39] The effect of this comparison, he suggests, involves the so-called "naturalization" of atomic danger and the engendering of a general belief that nuclear war might be both inevitable and survivable—just like natural disasters. In short, natural disasters in part taught Americans how to envision catastrophes of an atomic sort as it intersected with everyday life and its geography.

It is not surprising, then, to see a certain cartographic affinity between the two in their hypothetical imaginings. As Timothy Barney has persuasively argued, "a map's power is built on its ability to abstract and synthesize."[40] And Ann E. Larabee has detailed that the radial, utopian city, whose dense and productive engine core—"a nucleus, an informing *logos*"—outwardly fueled a concentrically ordered society, dominated imaginations of urbanity prior to 1945. But after Hiroshima and Nagasaki, the city core came to represent the center of risk: a Ground-Zero-to-come, a future nothingness at the center of radiating destruction.[41] Moreover, during the Cold War, insofar as security discourse was a play of simulations and projections, "since nuclear deterrence itself was a projection, the map provided a fitting vehicle for its presentation, allowing for the simple and reliable display of complex calculations and quantifications."[42] Specifically, the insertion of the atomic bomb into scientific disaster discourse encouraged and aligned what Matthew Farish has called a "spatial hierarchy of risk on the American landscape,"[43] a reconceptualization of the relationships between space, race, urbanity, population, and risk. It is well demonstrated in the literature on these cartographic simulations of atomic attacks that they were used to imagine the destruction of the urban core in major cities, the elimination of a racially-coded deviance used to justify white flight and the de-densification of the American population, just as often as they might have reflected something like a planning and response instrument.[44] They constituted a discourse of urban risk that articulated both the dangers of the city—Ground Zero—and the relative safety of positions outside of the radius of destruction. Their visual form, unsurprisingly, resembles the isohyetal rainfall bands of the SPS, but substitutes precipitation for annihilation.

For example, just a few years prior to the consolidation of the SPS, Ralph Lapp, a physicist who contributed to the Manhattan Project, wrote in the *Bulletin of the Atomic Scientists* in 1948 about the likely effects of an atomic blast targeting "an American city." Lapp's attention is primarily given towards the tactical elements influencing what an enemy would target and demarcating radii of effects, both physical and sociological. While the accelerating advancement of atomic technology would quickly render his assessment obsolete, its visual form remains instructive. In assessing something so elusive as "effects" of an atomic blast, and "for the purpose of simplifying the initial discussion," Lapp applies his analysis to a hypothetical scenario involving the targeting of a phantasmatic "City X." The city functions perfectly as a self-contained model: it remains general enough to stand in for a "typical American metropolis having a population of about 1 million people" and presumably important enough to be a target, but with features specific enough that an analysis of it could function as predictive, or at the very least instructive.[45] Lapp decides, for a host of strategic and psychological reasons, that the enemy will detonate this bomb directly over the center of the city at an altitude that ensures maximum destruction of both people and industry, hobbling the networks necessary for City X to respond in any meaningful way and inflicting the most devastating psychological effect on the nation possible. This approach is absolutely necessary for the experiment: the position of Ground Zero must be placed at precisely the most consequential place, for the effects of the simulation depend on this destruction. Lapp produces a map of the blast (Figure 3.5) that places Ground Zero at the center, with concentric circles radiating outward into zones at 1-mile intervals. Between these circles is another set that uses the severity of its zigzag pattern to communicate roughly the level of damage anticipated—the closest one being the sharpest and most angular, the farthest away patterning dispersal. The blast radius fits just perfectly into the city limits on two sides.

Lapp also briefly considers the effects of swapping Washington, D.C. for City X, with much of the same kind of analysis of where Ground Zero would be for maximum effect, how many would be killed, and what radius effects would result from the blast. The sequence is curious, given that both are make-believe: running the analysis of Washington, D.C. appears to serve as proof of the veracity of the principles detailed in the analysis of City X, rather than the other way around. While both scenarios are hypothetical, even if one is placed over an actually existing urban setting, the valence of authorization here recalls the era of physical modeling: newly wrought disaster scenarios provide the input for verification of the model. Nevertheless, this type of analysis was ubiquitous in the years immediately following 1945, including those produced by government entities like the Atomic Energy Commission in 1949. As Farish so persuasively argues, "what made such scenarios so chilling to American readers was not necessarily the gruesome description of the bomb's victims—since this is what [had been extensively] reported (however partially) from Japan—but rather the *location* of the destruction." Not just urban cores, dense with people, buildings, and industry, but *American* cities, and not just visual confrontation of the existence of such devastating technology, but the implication that it could be placed *here*, like a pin on a map.

But just as the very *raison d'être* of the maps can be traced directly to the atomic bombing of Hiroshima and Nagasaki and the mass anxiety it produced, the visual form of these maps is also traceable to official War Department discourse during this immediate aftermath. In June 1946, the Government Printing Office published a report as part of the post-WWII United States Strategic Bombing Survey entitled "The Effects of Atomic Bombs on Hiroshima and Nagasaki." Included in this report are aerial before-and-after images of each city.[46] In the Hiroshima pair (Figure 3.6), Ground Zero is marked by a reticle surrounded by a series of concentric rings spaced at 1,000-foot intervals. Rivers, roads, and the shadowy afterimages of a few buildings are all orienting features that prove, against incredulity, that the viewer is looking at the same geographic frame.[47] But the superimposed lines provide a visual logic of spatially radiating devastation, seen most

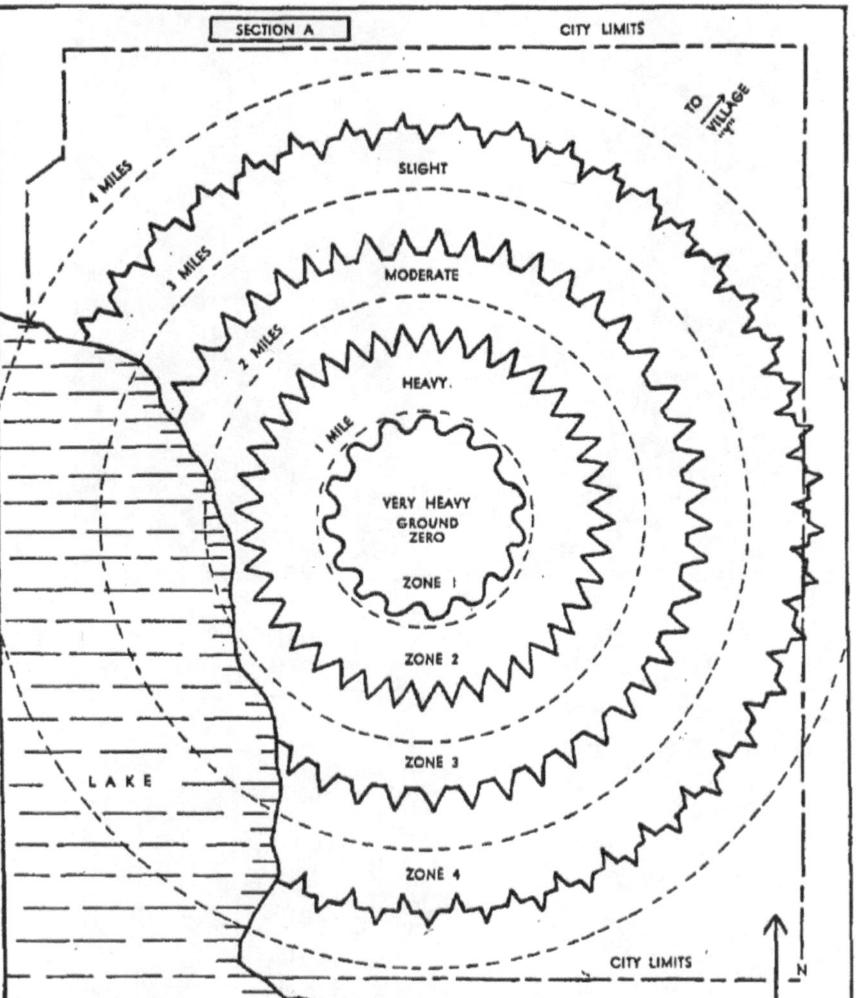

CITY "X"

Figure 3.5. Ralph Lapp's cartographic representation of the atomic bombing of "City X," 1948. *Source*: Permission to reprint granted by the Bulletin of the Atomic Scientists.

comprehensively in the report's fold-out inside back cover map of the extent of fire and structural damage (Figure 3.7).[48] The concentric lines, here also in 1,000-foot intervals, are supplemented by a red-shaded area marking fire damage and a pair of snaking lines tracing the mean- and maximum-distance areas of structural damage in Hiroshima.

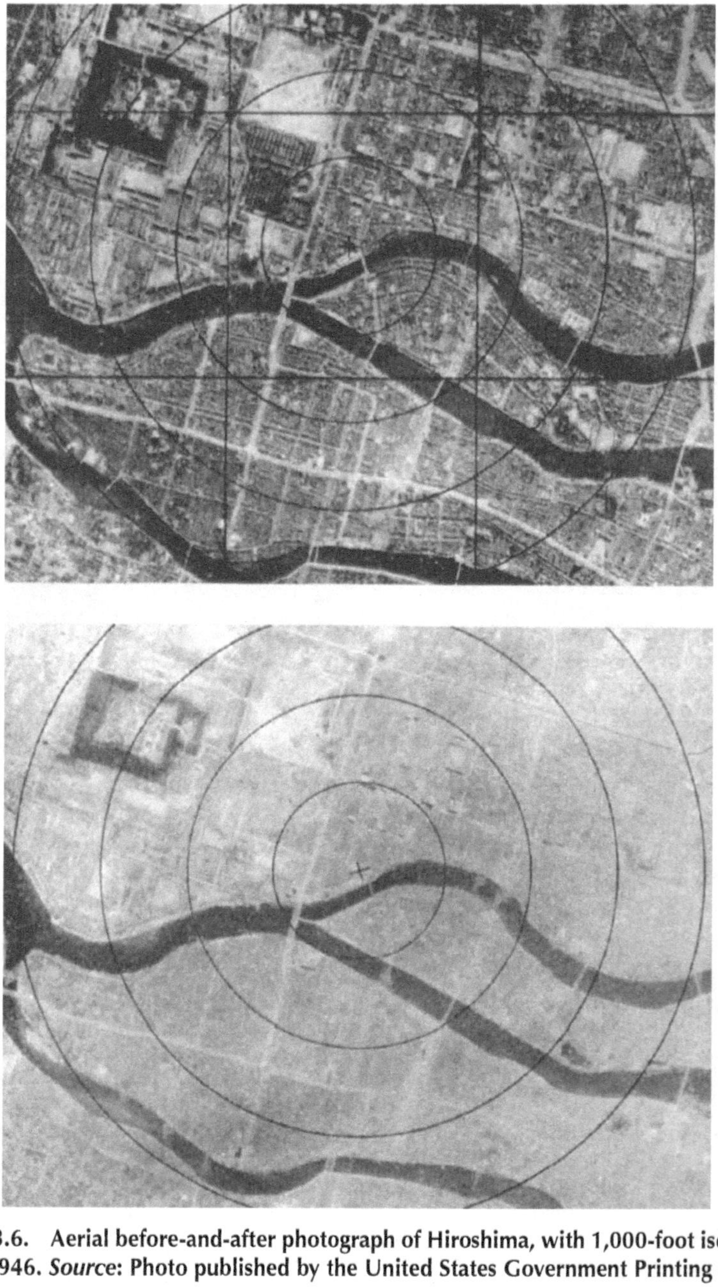

Figure 3.6. Aerial before-and-after photograph of Hiroshima, with 1,000-foot isohyetal rings, 1946. *Source*: Photo published by the United States Government Printing Office.

In both the hypothetical and forensic iterations of these maps, the concentric format of evenly-spaced concentric lines gives regularized and benchmarked structure to the true object of inquiry, which is not distance, but destruction, not geometry or geography, but the variable force of violence over space and its concentration at the center. As Larabee so emphatically argues, it is only from this perspective—in the bombing survey photographs, the bird's eye view, and in the maps, the vertical two-dimensionality of space—that one can imagine the sheer scale of death and destruction in geometric terms.[49] Through this form, the Hiroshima maps reconfigure space according to a geography of destruction: "the depiction of miles on these [types of] maps radicalizes distance as a vehicle of destruction; after a nuclear blast, the migration from the core to the periphery is a trail of death."[50] This, in the end, is the proper function of the visual form, and the isohyetal maps of the SPF do the same, concentrating the heaviest rainfall at the center of a regularized distribution of gradually lightening intensity which, at its outermost layer, ceases to be part of the storm at all.

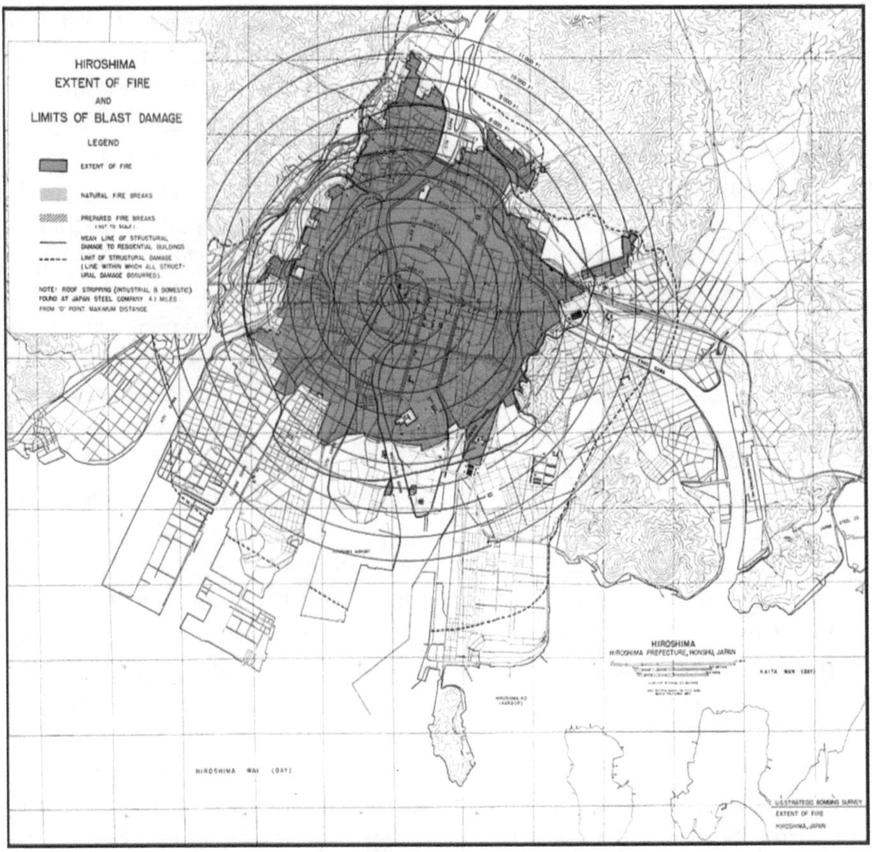

Figure 3.7. Map of fire and blast damage in Hiroshima, 1946. ***Source*****: Map published by the United States Government Printing Office.**

There is something entirely pragmatic, almost banal, about this coincidence of form, which is the simple, derivative fact that both storms and bombs have centers of intensity that gradually lessen outwards along a radius. As a relatively simple and ubiquitous visual form, the isometric geography of each could be dismissed as incidental if these visual forms had not achieved prominence in governmental and technical discourses of prediction and hypothesization at almost precisely the same time.

But another of the most striking effects of the inheritance of this visual format—that is, the visual logic of atomic blast cartography also in use within meteorological disaster cartography—is that storms could now be moved around like atomic bombs. Within the simulations produced by disaster predictors, the rippled, two-dimensional form of the Standard Project Storm is moved around like a pin on a map, placed often without explanation beyond anchoring a center point. The 1952 document, for instance, specifies only that "the 5,000 sq. mi. center area of each storm" were placed one over Clarksville, TN and the other centered over Burnside, KY, each standing in as the epicenter of the Lower and Upper Cumberland Basin, respectively. Involved in this is a two-part transposition subtly different from but aligned with Rheinberger's sequence: the first and most obvious being a geographic repositioning, the simple alteration in geographic spatial terms of the position of Ground Zero, and the second being the more technical definition as deployed by Army Corps scientists, which was the supplemental alteration of the data space to adjust for this repositioning on the basis of basin shape, average rainfall, and other factors deemed relevant to the model.

Transposition as a way of conceptualizing storms in federal meteorological discourse originated as early as the 1920s, and was only put into serious practice in the 1950s as standard project modeling came into a position of primacy, particularly for problems associated with extreme precipitation and flooding.[51] To extend this logic to hurricanes was only a matter of course. Not only was transposition a tried and tested way of splitting the difference between fabricating absent historical data and the lack of computational capacity to perform more accurate local simulations; as well, hurricanes share much with the tropological imagination of atomic bombs. Hurricanes are generated and deployed from abroad, they travel to the vulnerable shores of the U.S., and the relatively brief time between landfall and their dissipation or departure is effectively a period of detonation. A maelstrom rather than a conflagration, they nevertheless cause both immediate and long-term consequences in a widening, dissipating radius of force. This is a good enough metaphor for a particular hermeneutics of the Imaginary, but is hardly sufficient grounds for a meteorological science. Rather, what I hope to have shown in the past few pages is that a genealogical reading of the rhetorical forms used and reiterated by disaster scientists both atomic and natural in specialty demonstrates both technical and epistemological recursivity: natural disasters may have helped atomic scientists imagine nuclear disaster and its aftermath but, at precisely the same time, natural disaster modelers took up the rhetorical forms of civil defense cartography, rendering storms and the natural disasters they produce from within the formalizations of disaster brought to form by the atom bomb.

As the science of disaster prediction became increasingly enamored with statistical methods rather than quasi-empirical ones, and as atomic discourses came to provide the primary visual mode of understanding for mass destruction, the scientific conceptualization of disaster articulated the two halves: the mathematical shape of disaster became more regular, smoothed along and within the envelopes of intensity and recurrence, and were granted permission by cartographies of destruction to relocate, nearly intact.

In so doing, they transposed—displaced—public memory twice over. First, not entirely unlike the floods of the previous chapter, major storms were re-presented within the data space, abstracted from their material consequences, and this time abstracted from their original locations of effect. They were given new lives as hypothetical, malleable, and moveable mathematico-cartographic systems, used as ostensibly objective determinations of how much destruction is allowable in any given scenario, and consequently were themselves markers of allowance. That is, the project storms adopted for the purposes of design criteria were considered to lie at the intersection of the thresholds of allowable probability and intensity, which also meant drawing the line somewhere. This was made explicit in the technical discourse, for instance, in this excerpt within the 1979 SPH report:

> Statistical results when using limited data are subject to considerable uncertainty, particularly when developing values for rare recurrence intervals. Reliable observations have been taken for only about 80 years and there has been an average of less than one hurricane per year for the period of record of either coast. . . . Since the criteria must stand the test of time, meteorological judgment was applied to the few extreme events rather than relying heavily on statistical analysis.[52]

While the authors outline the procedure by which that "meteorological judgment" was rendered, the formalization has been articulated: project storms and their attendant models are a statistical endeavor, and where that statistical procedure becomes *methodologically* indefensible, supplemental calculations plug the gap on the basis of "expert judgment" in order to become *epistemically* defensible. Nevertheless, because project design and its economic calculus at this juncture had come to require the line smoothing—the upper and lower limits of some coagulation of reasonability and possibility—that data envelopes provided, and because their predictive capacity was bound inextricably to the statistical procedures that produced those smooth lines, the events of most consequence were the ones excluded from the model. Doing so, moreover, was the condition of possibility for the project storm's capacity for transposition, which tropologically and conceptually authorized design criteria as, if not necessary, at least reasonable.

And yet, while the conceptual move towards transposition had been made already, the development of the SPH in 1959 was followed by the devastation of Hurricane Betsy in 1965 and Hurricane Camille in 1969, two storms that rocked the Gulf region, caused billions of dollars of damage, and claimed together over 300 lives. The central irony of this exclusion has to do with the status of extremity in attempting to predict disaster. On the one hand, by dint of their position *as catastrophic*, Betsy, Camille, and other extreme hurricanes enveloped out of the project storm models because of statistical problems might have stood in—synecdochally—for precisely the type of storm one might design structures against. But the desire to pursue a different type of synecdoche, ironically, required the elimination of those storms from the methods of data reduction used to assemble the models.

This other, conceptually heterogeneous, version of synecdoche aligns more with Atilla Hallsby's deployment of the term in the context of political rhetoric. Following Ernesto Lacau and Diane Rubenstein,[53] Hallsby distinguishes between the traditional description of synecdoche—wherein a part of a whole stands in for the totality of the whole—and a psychoanalytic version that has a completely different ontological status:

> [S]ynecdoche does not substitute part for un unknown whole, but instead makes another w(hole) known by introducing an *additional* part. . . . Synecdoche, in other words, does not substitute part for whole. It is instead a part that *founds* the w(hole). Put in psychoanalytic terms, the *ontological* synecdoche is always registered as uncanny: it is an experience of social reality as simultaneously the same and not the same as it once was.[54]

This kind of synecdoche is one in which the *part*—for us, the model, the hypothetical project storm—is not only supplemental in the sense that it is not, in fact part, of the totality (all the storms that have been); it is also what *imparts* on the totality the coherence necessary for prediction to emerge as an epistemic endeavor. It is a storm that is not a storm, has never been a storm, and in many cases could not meteorologically exist as a storm, but that presumes to map the ostensibly finite terrain of the possible. Following this logical sequence, project storms also enabled both transposition and the cartographic visual homology—again, at an epistemic level—between storms and atom bombs by giving coherence, and thus existence, to the spectral events they imagined.

CONCLUSION

An exceedingly simple yet important fact, implicit throughout this chapter, authorizes this use of synecdoche and signals one of the next lines of inquiry: improved computational and technological capabilities are not the same as historical data. Climate risk scholars have argued precisely the same thing:

> In some studies, many years of record are simulated and statements are made which might lead the reader to believe that a commensurate gain in information over the existing historical record has somehow been achieved. However, a detailed review of these studies reveals the obvious fact that aleatory uncertainty [due to gaps in measurement data] cannot be overcome by simply executing numerical models over many randomly posed simulations, since the information upon which the parent distribution is based remains the same.[55]

And as these capabilities improved over the course of engineering science's advance into the computer age, including the extent to which simulated computer modeling was designed directly in reference to the improved quantity and quality of weather measurement technology, modeling past storms through simulation became a stand-in for unavailable historical data. The point is axiomatic to the entire edifice of this book, that simulations are technologies of both prediction and memory. The practice of simulating prior disasters was not a new one, as we have seen. But as computational capacity increased, simulations took on a new role. Whereas in the period of physical modeling, simulation was used for two purposes—verification processes were put into the service of proving the accuracy of the model by confirming the data that did exist, and then used to predict future disasters—the entrance into computational simulation gave rise to a third: using simulations to plug gaps in the data record. As computation became more sophisticated, the sheer volume of data that could be processed quickly and efficiently grew exponentially, and as instruments of weather measurement also became more sophisticated and provided orders of magnitude more data, the two became inseparable. The reduction of uncertainty (that is to say, the accuracy) that came with the ability to process ever more complex and voluminous parameters and variables began to be used as a rationale for backfilling the historical meteorological record, for attempting to deduce measurements that were never taken.

A second line of possible inquiry inverts the relation between public memory and its objects. In one part of the Nagasaki Peace Memorial Park, an obelisk stands at the center of a flattened, grassy area, planted at the hypocenter of the second atomic bomb's detonation—Ground Zero—and

Figure 3.8. Nagasaki Hypocenter, 2009. *Source*: Photo by James Gose, licensed under CC 2.0.

reaching skyward towards its epicenter some 500 meters above the ground (Figure 3.8). Radiating from this center obelisk are perfectly circular concentric rings of stone pavers, spaced evenly in larger and larger circles. Tangential to the largest of these circles stands a battered remnant of a wall belonging to the former Urakami Cathedral, a stunning Catholic church that was all but destroyed in the blast. The visual logic of radiating destruction is on full display here, seating firmly into the practices of memory the rhetorical forms most prominently used in cartographic and statistical modeling discourses. As it concerns the rhetoric of public memory objects, the inversion charted in the next chapter is fairly straightforward: instead of a reading method centered, from the start, on the form of the memory objects in question, these objects are taken as discourse effects of the memory work already occurring elsewhere—specifically, it should be no surprise, in technical and popular discourses of prediction and anticipation.

NOTES

1. Seed et al., "Investigation of the Performance of the New Orleans Flood Protection Systems in Hurricane Katrina on August 29, 2005," *Independent Levee Investigation Team: Final Report*, 2006, 12–13.

2. U.S. Congress, House of Representatives, *A Failure of Initiative: Final Report of the Select Bipartisan Committee to Investigate the Preparation for and Response to Hurricane Katrina*, 92.

3. Seed et al., "Investigation of the Performance," 12–16.

4. Seed et al., "Investigation of the Performance," 12–10.

5. "Blackstone River Flood Control Project at Woonsocket, Rhode Island," Technical Report No. 2-468 (Vicksburg, MS: U.S. Army Corps of Engineers, Waterways Experiments Station, 1957), 5.

6. Hans-Jörg Rheinberger, "Transpositions: From Traces through Data to Models and Simulations," in *Transpositions: Aesthetico-Epistemic Operators in Artistic Research*, ed. Michael Schwab (Leuven, Belgium: Leuven University Press, 2018), 217.

7. R. A. Fisher, "On the Mathematical Foundations of Theoretical Statistics," *Philosophical Transactions of the Royal Society of London: Series A, Containing Papers of a Mathematical or Physical Character* (1922): 311.

8. Fisher, "On the Mathematical Foundations," 311.

9. Fisher, "On the Mathematical Foundations," 322.

10. This is done according to a wealth of different methods, some of which are simply described in their specificity in this book, but it is beyond the scope of the current study to theorize them individually.

11. This concept is not entirely foreign to the practice of disaster modeling, even if Fisher radically transformed the way statisticians calculated error in his particular historical time. It will be remembered that model verification via simulation of historical data was a necessary part of legitimizing both the model and modeling itself during the period of physical modeling.

12. Donald T. Resio et al., "The Effect of Uncertainty on Estimates of Hurricane Surge Hazards," *Natural Hazards* (2013): 1444–45.

13. Donald T. Resio, Mark A. Tumeo, and Jennifer L. Irish, "Statistical Characterization of Hazards and Risk in Coastal Areas," in *Springer Handbook of Ocean Engineering*, eds. Manhar R. Dhanak and Nikolaos I. Xiros (New York: Springer, 2016), 571–72, emphasis in original.

14. *Handbook of Applied Hydrology: A Compendium of Water-resources Technology*, ed. Ven Te Chow (New York: McGraw-Hill Book Company, 1964), 9–60.

15. "Standard Project Flood Determinations," *Civil Works Engineer Bulletin* 52-8 (1952), 2–3.

16. National Hurricane Research Project, Howard E. Graham and Dwight E. Nunn, "Meteorological Considerations Pertinent to Standard Project Hurricane, Atlantic and Gulf Coasts of the United States," Report No. 33 (1959): 2.

17. National Hurricane Research Project, "Meteorological Considerations," 6.

18. Edgar Jadwin, "The Plan for Flood Control of the Mississippi River in Its Alluvial Valley: Presented to the Secretary of War and by Him to Congress," *The Annals of the American Academy of Political and Social Science* 135 (1928): 36.

19. Jadwin, "The Plan for Flood Control," 37.

20. "Standard Project Flood Determinations," 7. In later documents, it was referred to as the Probable Maximum Hurricane. See, for instance, U.S. Congress, House

of Representatives, *Interlying Areas Along Coastal Louisiana in the Vicinity of Houma: Letter from the Secretary of the Army* (1967), 22.

21. Scholars of weather science have critiques the apparent subjectivity of these definitions as well, either for their ambiguity or a lack of transparency about the data smoothing and fitting criteria. Resio et al. also point out that there are fatal assumptions within the premise itself, like the assumption that higher wind speeds will necessarily produce higher storm surges. Resio et al., "Statistical Characterization of Hazards," 576.

22. "Standard Project Flood Determinations," 10.

23. "Standard Project Flood Determinations," 24.

24. *Handbook of Applied Hydrology*, 9–62, my emphasis.

25. "Standard Project Flood Determinations," 26.

26. Chow, at the beginning of the *Handbook of Applied Hydrology*, calls the period between 1930–1950 the "period of rationalization" and argues that it was during this time that the mathematization of hydrology was accelerating into real usefulness with the proposal of many different calculations and coefficients. *Handbook of Applied Hydrology*, 1–9.

27. These were indexed as "depth-area-duration" calculations, which added to the precipitation and area calculations of the isohyetal maps the dimension of time, which was important for calculating the movement of water.

28. *Handbook of Applied Hydrology*, 9–62.

29. Rheinberger, "Transpositions," 215.

30. Rheinberger, "Transpositions," 219.

31. Rheinberger, "Transpositions," 220.

32. Given the centrality of psychoanalysis to this book, it should be noted that this conception of transposition is resolutely nowhere near Julia Kristeva's use of the term in her early work, which posited it as a reconfiguration of significatory relations themselves—something akin to a transformation of the Lacanian Symbolic. The present discourse under examination has its symptomatology, obviously, but transpositions described here are understood more squarely within the frame and methods of the Imaginary. See Kristeva, *Revolution in Poetic Language*.

33. Rheinberger, "Transpositions," 217, emphasis in original.

34. It is also not to say that one's definition of transposition cannot be rather capacious to be useful. Patrick Mahoney has argued persuasively, for instance, that Freud's deployment of *Übersetzung* expanded it conceptually well beyond its literal translations, including among its translations "hysterical, obsessional, and phobic symptomatology; dreams; screen memories; paraphraxes; the choice of suicidal means; the choice of fetish; the analyst's interpretations; and also the transposition of material from the unconscious to the conscious." Patrick Mahoney, "Toward the Understanding of Translation in Psychoanalysis," *Journal of the American Psychoanalytic Association* 28, no. 2 (1980): 473.

35. Resio et al., "Statistical Characterization of Hazards," 577

36. See also Resio et al., "The Effect of Uncertainty," 1444–45.

37. Although, it may be interesting to note, only between four-tenths and three percent of American citizens ever actually built or professionally installed private

shelters. See Kenneth D. Rose, *One Nation Underground: The Fallout Shelter in American Culture* (New York: New York University Press, 2001), 187; Sarah A. Lichtman, "Do-It-Yourself Security: Safety, Gender, and the Home Fallout Shelter in Cold War America," *Journal of Design History* 19, no. 1 (2006): 42.

38. Fred Charles Iklé, *The Social Impact of Bomb Destruction* (Norman: University of Oklahoma Press, 1958), 11.

39. David Monteyne, *Fallout Shelter: Designing for Defense in the Cold War* (Minneapolis: University of Minnesota Press, 2011), 1.

40. Timothy Barney, *Mapping the Cold War: Cartography and the Framing of America's International Power* (Chapel Hill: University of North Carolina Press, 2015), 9.

41. Ann E. Larabee, "Ground Zero: The City, The Bomb, and the End of History," *Canadian Review of American Studies* 22, no. 2 (1991): 267.

42. Barney, *Mapping the Cold War*, 178.

43. Matthew Farish, "Disaster and Decentralization: American Cities and the Cold War," *Cultural Geographies* 10, no. 2 (2003): 126.

44. In addition to other cited in this chapter, for further insight see also Laura McInaney, *Civil Defense Begins at Home: Militarization Meets Everyday Life in the Fifties* (Princeton: Princeton University Press, 2000), 41.

45. R. E. Lapp, "Atomic Bomb Explosions: Effects on an American City," *Bulletin of the Atomic Scientists* 4, no. 2 (1948): 49.

46. The United States Government Printing Office, "The United States Strategic Bombing Survey: The Effects of Atomic Bombs on Hiroshima and Nagasaki, June 30, 1946," (1946), 7, 10.

47. See also the metonymic comparison of these two images with an ancient biblical depiction of the walled city of Jericho in Paul Virilio, *War and Cinema: The Logistics of Perception* (New York: Verso, 1989), 70.

48. The United States Government Printing Office, "United States Strategic Bombing Survey," inside back cover.

49. Larabee, "Ground Zero," 268.

50. Barney, *Mapping the Cold War*, 203.

51. *The Handbook of Applied Hydrology*, 9–60

52. U.S. Department of Commerce, National Oceanic and Atmospheric Administration, "Revised Standard Project Hurricane Criteria for the Atlantic and Gulf Coasts of the United States," Memorandum HUR 7-120 (1972), 135.

53. Ernesto Laclau, *On Populist Reason* (London: Verso, 2005); Diane Rubenstein, *This Is Not a President: Sense, Nonsense, and the American Political Imaginary* (New York: New York University Press, 2008).

54. Atilla Hallsby, "Imagine There's No President: The Rhetorical Secret and the Exposure of Valerie Plame," *Quarterly Journal of Speech* 101, no. 2 (2015): 363–64.

55. Resio et al., "Statistical Characterization of Hazards," 570.

Chapter 4

Prefiguring Hurricane Katrina and the Rhetoric of "The Big One"

INTRODUCTION

At the end of the previous chapter, I signaled two possible lines of inquiry that could be derived from the implications of the statistical and cartographic Cold War-era of disaster modeling.[1] First was the relationship between simulation and history: could simulations be used, serviceably if not entirely truthfully, to plug the historical record gap? Like astronomers in ancient times who, having calculated the trajectories of the stars, were able to past-date eclipses, could improved computational technology be used to simulate past storms for which the meteorological record was incomplete in order to retroactively construct their details? As we've seen argued, though one could absolutely construct such a fabrication, there is really no epistemological justification for it. Or, at the very least according to modern scholars, because the source data remains unchanged, plugging the historical record cannot be used as a substitute for a sufficient quantity of existent data. Thus, one could use statistical modeling to approximate, say, storm surge levels or wind speech during the famous 1900 Galveston hurricane, but to do so would be more novelty than, strictly speaking, educational or informative. And it is certainly the case that one could not then treat those approximations as source data for model verification for the reasons cited.

Yet, despite the fact that historical data have not changed, developments in modeling and simulation technology have improved disaster forecasting, both in the long-term calculation of risk and in real-time tracking and response to storms. This development alters the discourse on the science of prediction, particularly, for us, the relationship between prediction and public memory: if disasters are by definition transgressions, what does it mean to collectively process a disaster event that everyone saw coming? An event whose parameters, if not precisely its effects, were essentially known in advance? In this chapter, we arrive, finally, at Hurricane Katrina.

THE BIG ONE

In June 2002, a widely distributed fictional account of what post-hurricane New Orleans would look like painted a picture of massive flooding, tens of thousands of deceased, infrastructure obliterated, and months upon months of efforts to "drain the bowl" that is the city's geography. Using advanced computer modeling that "updated" forty-year-old data, the narrative describes a near-apocalyptic scenario of a Category 4 or 5 hurricane in uncomfortably specific detail. It would be "at least four days after the hurricane" before crews could begin to pump water out of the city, an endeavor further complicated by the fact that existing pump infrastructure would most likely be useless and broken by the storm. Replacement pumps would "take six months to pump out Jefferson Parish" and even longer to drain the entire city. That timeframe would compromise sanitary systems and the integrity of buildings and infrastructure, and would likely spark potential fires once electricity was restored. Refugee centers and temporary residences at military bases would be established. Countless businesses and individuals would permanently relocate. Although it is unlikely that the city would be abandoned altogether, the narrative admits, it is hard to overestimate how devastatingly such a storm would change New Orleans. The exigency for telling such a story is then revealed: "It's a matter of when, not if."[2]

Two years later, in July and August 2004, another similar story, but with exponentially more detail. "Hurricane Pam," a fictional, slow-moving Category 3 hurricane, makes a direct hit on New Orleans, "leaving in its wake 61,000 fatalities"[3] and nearly 400,000 nonfatal casualties despite the more than 1,000,000 evacuated residents. Projected realities include a "HAZMAT 'gumbo'" of toxic and possibly flammable hazardous chemicals,[4] millions lacking "power, water, and ice,"[5] transportation nightmares for 500,000 people needing rescue and/or relocation,[6] an unprecedented need for volunteers and donations,[7] a vetting process for re-entry to the city once safe,[8] the removal of "30 million total cubic yards of debris" (including coffins but not including human remains),[9] 250,000 displaced schoolchildren,[10] 500,000 people requiring post-landfall shelter[11] and/or temporary housing,[12] and inadequate medical capabilities.[13] In other words, a natural disaster for which existing and projected capabilities were determined in advance to be not just inadequate, but disastrously so.

These two scenarios each figure a strikingly similar sense of anticipation. The first narrative was the product of a five-day special report entitled "Washing Away" published in 2002 by the *Times-Picayune*, New Orleans's largest newspaper, based on new scientific analytics and projections that it had commissioned. Drawing on these projections, as well as the history of natural disasters in the region and more recent natural disasters in other regions, the report paints a picture of what would happen—indeed, what *will* happen, for it uses the simple future tense throughout—when "The Big One" hits. Its warnings are rooted most centrally in the deadly combination of a lagging disaster infrastructure and accelerating urban development. The second narrative is the Southeast Louisiana Catastrophic Hurricane (SLCH) Functional Plan generated by Innovative Emergency Management, Inc. (IEM), commissioned by FEMA and the Louisiana Office of Homeland Security,[14] and produced during the weeklong SLCH exercise in July 2004. IEM was the primary contractor during the SLCH exercise tasked with running a simulation of a slow-moving Category 3 hurricane, which it named "Hurricane Pam."[15] IEM published its 121-page report roughly two weeks later, which detailed the expected effects of such a storm, what was required to respond to and manage those effects, and where capabilities across federal, state, and local levels were likely to fall short. Although the report takes a far less apocalyptic tone than the *Times-Picayune* special report, it nevertheless constructs a narrative of post-disaster Southeast Louisiana that is by implication just as catastrophic.

These are but two of the more recent and widely visible iterations of a longstanding regional trend towards narrativizing The Big One, a singularly destructive hurricane so devastating to New Orleans and the surrounding areas that it takes tens or hundreds of thousands of lives, spurs a mass migration of people and industry, swallows the city whole, and/or turns it into something else entirely. "The Big One" is a phrase associated with other assumedly imminent disasters throughout the world as well, including a massive earthquake along the San Andreas fault,[16] another earthquake farther north along the Cascadian Subduction Zone,[17] particularly devastating tsunamis in Southeast Asia,[18] and the Yellowstone supervolcano eruption.[19] These events and others are similarly anticipated, endlessly rehearsed, and form the basis for an anticipatory mourning of a tragedy that will at some point have happened.

Whereas the object foci of the previous two chapters were by and large the technics of and debate surrounding the scientific discourses themselves, this chapter looks much more closely at how widespread public perceptions of prediction—figured through the signifier "The Big One" and authorized by the simulations that meteorological and geological sciences provided—prefigure the rhetorical forms that public memory objects take on. In other words, simulation in these cases not only traffics the memory of disaster through its data via functionally similar means as those of the past, but now also, by dint of its improved accuracy and therefore the believability of its forms, primes the invention of memory discourse proper. The theoretical question raised in Chapter 2 about the emergence of memory objects finds a theoretical answer here, through the Freudian concepts of melancholy and incorporation, by tracing two such simulative narratives from prediction, through event, and on to memory.

While the regional particulars of New Orleans's Big One populate the bulk of this chapter, the anxiety structure constitutive of this anticipatory mourning and its articulation to a more broadly disastrous imaginary domain need to be situated both conceptually and historically. In tracing what amounts to a genealogy of anxious mourning in New Orleans as I do in the next section, one senses a more durable historical and rhetorical structure of anxiety to which Freudian thought helps give language, as well as the more important rhetorical configuration of anxiety and melancholy—anticipation and memory—in the present age of heightened and increasingly ubiquitous disaster. The Big One, while consistently figured as a natural disaster, only occupies this singular figural position by dint of its uncontrollability. That is, without irony, stories of The Big One repeatedly imagine the apotheosis of the failure of control by attempting to prefigure it, to lock it into place in advance by way of rhetoric.

One upshot of this for rhetorical studies of public memory in the postmodern era is an addendum to the conversation about the emergence and constitution of public memory's objects, for two reasons. First, what is involved in such narratives is the memorialization of futurity or, more precisely, the future anteriority of subjects adjacent to whom events will have happened. Barbara A. Biesecker argues that the melancholia of this tense "may not be understood simply as either a reaction to a historical loss or the playing out of a structural lack, since the loss is the ghostly predication of the subject who intends to mourn."[20] In the current age, replete with disaster but absent both the strict eschatology of the pre-Enlightenment era and the positive teleology of modernity, assumed in advance is the inevitability not only of a disaster to be mourned, but also of a collective subject who does the mourning. Within this space, positioned between anticipation and loss, can be found anxious melancholic rhetoric, whose manifestation figures the event so repetitively over time that one must reasonably presume that any public memorialization emergent from the aftermath of actually-occurring disaster could not avoid incorporating these figurations at some level. This assumption takes for granted the challenge that Carole Blair, Greg Dickinson, and Brian Ott issue to the literature on remembering and forgetting, a dialectic which they argue "has been employed as a stand-in or simplistic restatement of the problem of representation in public memory studies."[21] Alongside Bradford Vivian's work on public forgetting,[22] most useful about their critique of the tendency of scholars to align memory and remembering, set against forgetting, is its formal articulation to another dialectic of presence and absence: "a failure to represent a particular content publicly is not a necessary, or even provisional sign of forgetting."[23] Rather, they encourage scholars to consider the particularities of representation, figuration, and memory in attempting to describe the work of public memorialization. This chapter's response to that challenge attempts to do so by tracing the melancholic incorporation of the anxieties of imagined futures into objects of public memory.

Second, natural disasters today always bear the mark of anthropogenic climate degradation in both its ecological and social forms, articulating yet another layer of human failure and proposing yet more opportunity for even greater disasters. In other words, Huet's thesis that disaster and human failure to a significant degree organized Western philosophy and "durably transform[ed] the post-Revolutionary age,"[24] explored in Chapter 2, is further amplified in an era in which human failure extends from the irreversible destruction of redlined Black neighborhoods in the South to the violence of the hurricane that wrought it, in its very existence and its intensity and not merely in the fallout of its incidental passing. By extension, so too does the expectation of memorialization in the aftermath of such failure.

New Orleans, as a place historically known both for its environmental vulnerability and for its troubled relationship to these Western ideals, serves as a crucial object lesson within the rationality of climate crisis and late stage capitalism, when hurricanes and floods are intensifying at a rate with which crumbling disaster response and mitigation infrastructure—both physical and administrative—have not kept pace. The special kind of environmental and social fatalism that infuses talk of The Big One produces a rhetoric instructive for an increasingly likewise fatalistic set of discourses about the general inevitability of disasters to come, and subsequently about how they will be remembered. For Katrina, then, critically reading its public memory objects is not merely a matter of beginning at the exact moment of the storm, detailing the event itself and then the rhetorical practices whereby the event is remembered in certain ways and forgotten in others, and then deriving conclusions on the basis of how Katrina is and is not remembered in an archive which bears its name. Instead, I argue, that the repeated narrativization of The Big One makes up an anxious rhetoric that, in staging disaster, thereby prefigures post-Katrina memory objects through a process of melancholic incorporation. I first engage a history of New Orleans and this anxiety, extrapolating the relationship between anxiety and melancholy along the way. Doing so recruits a regional narrative into the service of laying out a rhetorical reading strategy based on this Freudian articulation. Following that, I conduct a two-part reading of a rhetorical sequence: first, I attend to the anxious rhetoric of the two narratives of destruction that opened the present chapter to trace this prefiguration, and then I turn to a reading of post-Katrina memory objects to demonstrate the work of melancholic incorporation in their production.

HISTORY, ANXIETY, AND MELANCHOLY IN NEW ORLEANS

At the broadest level, the history of New Orleans is the somewhat generic tale of the colonial desire to claim and tame the environment within a profitable and pragmatic plot of land. On one side, the city is merely the southernmost location of myriad struggles to control the Mississippi over the course of the last 300 years. The enormity of the river and the tendency of that enormity to translate into similarly scaled flooding, the need to build adequate bridges over it, its central role in economics, trade, and development, and the task of facilitating those activities by modifying the river itself: these imperatives and others have straddled the line between hubris and opportunity, between overextending the capability to manage the river and taking advantage of the rewards of development and civilization that its unique qualities provide. On another side, butted up against Lake Pontchartrain, New Orleans has a similarly storied history of vulnerability to the effects of major hurricanes, often exacerbated by the warming waters of the Gulf of Mexico and those very changes to the Mississippi riverscape that development has brought. The major threats posed by hurricanes, as was clearly evident during Katrina, come less from the massive winds and rain than from storm surges and river swelling, both of which have the potential to cause catastrophic flooding and infrastructural damage. Because of this dual threat, development of New Orleans since its first founding has been an exercise in engineering, from a vast system of levees, to artificial channels, to houses built on stilts, to complicated pumping stations designed to keep the land dry from flooding. From this perspective, New Orleans is, and has always been, a city wrestling for stability within the environment that sustains it, a place where catastrophe is said to map directly onto its urban architectural logic. It is a narrative full of both allure and danger (or desire and disaster, as Lynnell Thomas puts it),[25] not so different from those about the city itself, and one that contemporary scholars continue to tell. One particular retelling of this story, Christopher Morris's *The Big Muddy* (first cited in Chapter 3), furnishes an opportunity to offer a parallel reading: alongside a critical summary of this narrative of colonial and environmental mastery, one can simultaneously chart the evolution of Freud's thinking on the question of anxiety and the opening out of its discourse onto the domain of the melancholic.

In a thoroughgoing environmental history of the Mississippi Valley that comprehensively threads through Native American, Spanish and French colonial, and modern historical eras up to Katrina, Morris tells the story of the constant struggle for dry, stable land. He argues that Mississippi Valley living is best understood within a dichotomous frame of wetness and dryness: "There are two Mississippi Valleys. One is wet, the other dry. The river made the wet valley by flooding it with dirty water and filling it with mud into dirt. The two valleys exist in uneasy tension, the wet valley always ready to burst into the dry valley that holds it down."[26] The juxtaposition of wet and dry is posited here as being constitutive of the Valley itself, an inexorable coagulation of two extreme conditions of (de)hydration. In peaceful times, the river sleeps within the confines of its banks, and plant and animal life thrive around it unperturbed. In more violent ones, the river awakens, overtops its banks, and once more consumes the dry land indiscriminately, only to leave behind an even greater quantity of sediment. Dry land here owes its existence to the river that deposited it. At the same time, however, it is also always under constant anticipation of inundation by that very same force, as it has continuously throughout centuries of settlement and development.

Altogether, the central political claim of Morris's narrative is that this gradual and steady drying-out of New Orleans and the Mississippi Delta produced a host of environmental and social conditions whose logical conclusion was Hurricane Katrina. The technologization of the land and water have caused erosion that has decimated the coastline, natural marshlands that need seasonal flooding to exist have receded (intensifying the impact of storm surges), and the drying and compacting of the land is causing the city, slowly but surely, to sink farther below sea level. In the early eighteenth century, these measures prevented drainage when water *did* enter, stagnating into pools of sickness and decay. Today, technologization of the land has allowed development to spread into areas even more vulnerable to disaster and, as Katrina roundly demonstrated, these technologies have not kept pace with the very threats that they have intensified. This philosophy of development has appeared to paint New Orleans into a corner, rendering such a philosophy an eco-technological *pharmakon*: a cure to the city's ailments that is also poisonous to its own aims.[27]

Morris's recommended response to this dilemma is curative: allow nature to have some of its nature back; learn to live with the water, not against it. Since the city clearly cannot keep the Mississippi hypersomnic, and since New Orleans's struggle to shield itself from the river exacerbates the river's effects, the answer is to allow The Big Muddy its seasonal release, but in an anticipated and controlled way. In Morris's final estimation,

> If New Orleans is to be sustained, it will be through a combination of water manipulation and water acceptance. Nature always takes its course. Repressed in one place, it returns somewhere else. City blocks abandoned since Katrina are rapidly reverting to impenetrable jungles like those [La Louisiane founder Pierre Le Moyne d']Iberville described over three centuries ago, thickets of saplings and vines harboring snakes, alligators, and small mammals, and shrouding the ruins of a city built in denial of nature.[28]

This assessment is nothing if not an analogue to psychoanalytic "working through," and its logic is Freudian throughout. "Repressed in one place, it returns somewhere else," "a city built in denial of nature," "*nature always takes its course*." No matter how effectively the subject represses, the trauma of the unconscious always bursts forth. The key for Morris is to develop a strategy whereby the river's diluvial outbursts are acknowledged and lived with, rather than against. Morris's diagnosis of denial, the Freudian *Verleugnung*, is more accurately disavowal, and displaces the anxiety over the city's dryness onto the technologies that purport to keep it dry.

However, a more careful reading of the book's own structure of anxiety reveals something absolutely fundamental to the framing of the city's dilemma. We may recall the book's opening sentences: "There are two Mississippi Valleys. One is wet, the other dry. . . . The two valleys exist in uneasy tension, the wet valley always ready to burst into the dry valley that holds it down." Here we have a description of the Valley in its "nature," absent humans and yet—still, there is an uneasy tension. One might surmise on the basis of Morris's own curative conclusion that there is in fact nothing "tense" about its existence at all: it is not only perfectly "natural," indeed, it is constitutive of the Valley itself.

Allowing the river to serve provisionally as a useful metaphor for the unconscious, this contradiction invites a brief consideration of the evolution of Freud's theorization of anxiety. His early essay on anxiety and phobia situates the former as an effect of what he calls "unemployed libido," the result of the ego's attempts to shield itself against primary anxieties developed in childhood:

> In phobias, for instance, two phases of the neurotic process can be clearly distinguished. The first is concerned with repression and the changing of libido into anxiety, which is then bound to an external danger. The second consists in the erection of all the precautions and guarantees by means of which any contact can be avoided with this danger, treated as it is like an external thing. Repression corresponds to an attempt at flight by the ego from libido which is felt as a danger. A phobia may be compared to an entrenchment against an external danger which now represents the dreaded libido.[29]

In other words, repressive mechanisms "dam up"[30] the ideational content of dangerous libidinal energy, which is then discharged as anxiety directed at a phobic object in what Sarah Beardsworth calls "an economic process consequent on repression."[31] Repression first, then anxiety. Later in his career, however, Freud places anxiety more squarely in the everyday workings of the psyche. As Beardsworth notes, within this revised framework, Freud theorizes anxiety on the basis of ideational content rather than libidinal energy which leads him to the conclusion that "it is not repression that creates anxiety but anxiety that makes the repression."[32] Faced with the perception of the threat of danger, anxiety-as-signal activates various defenses that work to shield the ego from that danger. This process comports with both Morris's description of the central problem that New Orleans faces and his prescription to eliminate the hydraulic economy on which it is founded. More importantly, it allows the critic to read anxiety as a generalized function of discourse rather than as only the product of individualized repression: the anticipatory rehearsal of the threat itself is what builds the psychic levees. Generalized as rhetorical function rather than psychic product, the rhetorical manifestations of this anxiety for New Orleans consistently take the form of a Big One that posits the danger as not only inevitable, but also definitively external.

Such repetition invites a critical reading of melancholy adjacent to but distinguishable from that implicated by Robert Hariman and John Louis Lucaites's "archaeology of the present," imagistic figurations "that remind us of how close we might be to becoming ruins." For them, the disintegration of modernity produces catastrophe as an historical horizon, and images of ruin reveal to viewers not only how close we may be to ruin, but also that "potential demise is already coded into the present, already one possible future extending from the present."[33] Figurations of ruin serve as allegory for the fall of modernity, evidence that whatever fortifications exist that mitigate against the cruelty of a so-called state of nature are as fragile as they are arbitrary. The aesthetic or, in Benjaminian fashion, "attitude" of such allegory for Hariman and Lucaites is melancholy, "just as the ruin is its central motif." "The ruin is a fragment," they write, "a trace, a sign of time's corrosiveness, a thing reduced to being an accidental sign, a call to memory, and more as well."[34] In effect, the ruin is an image symptomatic of an historical attitude that has jettisoned the pretenses of Enlightenment progress, and the resultant mood is the mournful gloom of the melancholic. If Huet's diagnosis of the originary philosophical problem underwriting modernity is to be believed, however, this attitude is the manifestation of a rhetorical trick: because modernity is founded upon the dialectical anxiety of containment and disaster, it is quite appropriate to think of the repetitious figuration of disaster as melancholic not just in its mood or aesthetic, but also in its structure.[35] That is, the rhetorical reading strategy that I am suggesting attempts to first locate an anxious form and its later transposition that often, though not always, results in what is normally recognized as a melancholic aesthetic.

The basic process that details this progression Freud calls incorporation. As Judith Butler explains in *The Psychic Life of Power*, melancholy, "the unfinished process of grieving,"[36] involves the incorporation of the lost object in the development of the ego. The loss of a loved object inaugurates a grieving process whereby the subject must somehow "let the object go." Freud, she argues, came to see the process of grief not through the detachment of a subject from the loved object but, rather, "he makes room for the notion that melancholic identification may be a *prerequisite* for letting the object go. . . . [changing] what it means to 'let an object go.'"[37] Rather than becoming detached, which would represent a successful completion of mourning, the object is instead incorporated into the ego through identification with it:

> The lost object is, in that sense, made coextensive with the ego itself. Indeed, one might conclude that melancholic identification permits the loss of the object in the external world precisely because it provides a way to *preserve* the object as part of the ego and, hence, to avert the loss as complete loss. . . . not full abandonment of the object but transferring the status of the object from external to internal.[38]

Melancholy is therefore not simply unfinished grieving and should be thought outside of the specific act of grief. The introjection of the abandoned object-cathexes, Freud argues, may well precipitate the character of the ego itself, which "contains the history of those object-choices."[39] To most simply describe this structural relationship between anxiety and melancholy: the traces of all traumatic loss, both anticipated and experienced, etch themselves into the ego by transforming into more acceptable, manageable symbolic forms.

To sum up, the anxious rhetoric that repetitiously anticipates, by rhetorically defining and delimiting the event in advance, and therefore pre-mourns disaster functions by incorporating the objects of this anxiety into the practices of public memory. Analysis of Katrina's public memory thus requires the mapping of this melancholic process to describe fully the labor that post-Katrina public memory practices perform. While Katrina was in many ways The Big One that residents and experts knew would some day come, for New Orleans there will always be another, that sweeping blow that will finally (retroactively) be recognized as having been The Big One. Before that, however, of necessity is a deeper reading of two of these narratives that prescribe and organize its parameters.

Hurricane Pam

On January 14, 2006, mere months after Katrina had battered New Orleans, flooded the bowl, and ultimately, without adequate help from the federal government, left many of its citizens either dead or indefinitely evacuated, the Committee on Homeland Security and Governmental Affairs (HSGAC) held a hearing to assess the role that the fictional Hurricane Pam exercise had played in the preparations for a catastrophic hurricane in New Orleans. Unlike the September 22, 2005 hearing mentioned above, this hearing did not investigate the advanced warning for Katrina itself; instead, its primary concern was the role of The Big One in disaster preparedness and response. In the opening statements to that hearing, Senator Susan Collins (R-ME) makes a telling analogy: "As a dry run for the real thing, Pam should have been a wakeup call that could not be ignored. Instead, it seems that a more appropriate name for Pam would have been Cassandra, the mythical prophet who warned of disasters but whom no one really believed."[40] The basis for the analogy is that, despite the uncanny similarities between the details of the simulation and the details of the actual storm, the disastrous governmental response from local to federal levels and the actual effects of Hurricane Katrina mirrored the projected effects of Hurricane Pam in nearly every way except fatalities. If, she implies, the purpose of Pam was to predict the tragedy precisely in order to plan for and mitigate against its consequences, and given that the conditions of both were so strikingly similar, why, then, were those effects still so disastrous? At stake in this analogy is that, beyond the minor irony of the fact that unlike Cassandra everyone believed Pam, it produces an inverted logical sequence: first, believability is endorsed through appeals to realism; second, this sense of realism is posited by way of a constitutive exception; and finally, this exception becomes an unspoken condition of success or failure, retroactively organizing what constitutes a "realistic" or "believable" disaster scenario.

Hurricane Pam endorsed a notion of believability by first inverting the normal procedure of disaster planning. In one of the most symptomatic portions of the hearing on that day, both Sean Fontenot, a Louisiana State preparedness official, and Madhu Beriwal, IEM president and CEO, describe the uniqueness of Pam in this regard. As Fontenot notes, "From the word 'go,' it was understood that this was not a typical exercise. . . . Usually, you write a plan and then have an exercise."[41] Pam was the direct opposite: a specific scenario was posited, and then a plan was devised in accordance with that scenario. Beriwal, elaborating on the rationale for this approach, testifies that "[t]he intent of Hurricane Pam was to create a plan for a catastrophic event, a specific event . . . the intent was to create a sense of reality. When we were working with this project, we were trying to describe a worst case but plausible event. That is the slogan that we had."[42] A FEMA news release published directly after the exercise confirms this interpretation.[43] Recalling Jacques Derrida's nuclear-age insistence that deterrence, for all its reliance on the technological means of mutually assured destruction, also relies upon both the rhetorical performance of belief in that outcome and "the rhetorical simulation of a text,"[44] one of the rhetorical functions of Hurricane Pam was to perform believability through an appeal to realism, as well as to produce a simulation in the hope that it might act as a substitute. "Instead," Collins laments, "Pam became Katrina. The simulation became reality. And optimism became the awful truth. We were not prepared."[45] Katrina could never have been other than, in the words of Senator Joe Lieberman (D-CT), "the storm people in the Gulf Coast had always feared, the storm people knew would hit one day, the storm they actually practiced for in the Hurricane Pam exercise that is the topic of today's hearing."[46] By implication, part of Gulf Coast existence involves living within that logic of not if, but when.

What is remarkable about the Hurricane Pam exercise, however, is that its entire framework is organized around an unquestioned assumption that the levees would hold. The details of this assumption are telling. Just days after the levees broke, the chief of engineers for the U.S. Army Corps. of Engineers denied a failure on the part of the existing levees. In short, he argued, the levees were always designed for a Category 3 hurricane, not a Category 4 or 5. "It was fully recognized by officials," he said, "that we had Category Three level of protection. As projections of Category Four and Five were made, [officials] began plans to evacuate the city."[47] Pam may well have simulated a Category 3 precisely to operate under that assumption. And yet, despite the fact that Katrina did not technically hit New Orleans directly, despite the fact that its force matched almost exactly the projections that Pam laid out, and despite the fact that Katrina was actually a Category 3 at the time of landfall, this argument inexplicably takes the most central assumption of Pam as its locus of deniability. Indeed, the very first page of the SLCH report, which begins with a section on "unwatering" the bowl, makes clear the premise: "For the purpose of this plan it is assumed there are no levee breeches [*sic*]. This is the worst case situation." Furthermore, "much other response and recovery activity depends on the successful unwatering of bowls, at least to the +2[-]foot elevation."[48] By implication, failure to unwater the bowl—and, in the case of Katrina, the massive underestimation of the scale of the bowl's watering—would render the rest of the report of limited practical value. Thus, what is most remarkable about Pam is not simply that it dismisses the possibility of levee failure, but rather that this dismissal is the condition on which all the details of effective response specifically rely.

As an exception, "the element with no place in the structure,"[49] levee failure is the central anxiety that animates and organizes the entire SLCH Functional

Plan and thus its presumptions of realism and believability. In beginning with a catastrophe and then developing a plan, rather than the more typical inverse, Hurricane Pam did not so much simulate the potential realities of The Big One; instead, it organized and limited the parameters of the event in advance. As FEMA Response Operation Branch Chief Wayne Fairley testified, "The project did not result in a catastrophic planning document per se, but rather a *framework* for developing such a plan."[50] It attempted to freeze the event within a set of conditions that could be planned against but, in so doing, inadvertently named the exception without which those plans fell apart.[51] It is only in this sense that Senator Collins is correct in saying that "Pam became Katrina," that "simulation became reality." If Pam was the "realistic" rhetorical simulation of The Big One that its inverted process purported it to be, then the actualization of the fantasy of The Big One would require the very contingency of failure that it dismissed. Homeland Security Official Jesse St. Amant recognized this when he testified, "You build a 20-foot levee, Mother Nature will give you a 25-foot storm surge. . . . [W]e can blame everything and his brother for what has happened, but the fact of the matter is . . . we knew in this business that this was coming. We tried to say the words, this is coming, time and again."[52]

Washing Away

The Times-Picayune's "Washing Away" series takes this assumption to heart, describing in vivid detail what such failure would look like. Unlike Pam, whose object of anxious projection was levee failure as an exception, "Washing Away" finds its object in the desire to witness failure and the erasure of human presence directly through an act of imagination. The first installment, entitled "In Harm's Way," frames the entire series by first recounting two major hurricanes in 1856 and 1909 and articulating them to Louisianan identity. Both of these storms, it tells us, impacted Louisiana resident Claire Rose Champagne's ancestors, sometimes at the cost of their lives. Although the family moved 30 miles inland after the first hurricane, "there was no escape from the storms, which have followed the family inland for over five generations." Even today, Claire and her family have accepted that life in Southern Louisiana is in no small part a life governed by the rhythm of seasonal hurricanes. "Hurricanes are a common heritage for Louisiana residents," the introduction assesses, "who until the past few decades had little choice in facing a hurricane but to ride it out and pray." The report then systematically lays out the dilemma that Morris and others have also outlined: The development of technologies that provide protections against these storms have made the coast "more vulnerable to hurricanes, not less."[53] As it progresses, the scope of this section quickly narrows to New Orleans itself, and for the remainder of the special report never quite leaves. Whatever larger effects development has had on the Gulf Coast as a whole, those effects are profoundly amplified when it comes to New Orleans.

Rather than dismissing levee failure, the purpose of "Washing Away" is precisely not only to assume failure of all kinds and imagine subsequently disastrous effects, but also to attempt to imagine a "megadisaster" in New Orleans whose scale would be bigger than anything the nation had yet seen. In attempting this, it repeatedly places touchstones of disaster and failure in both New Orleans and throughout the world. For instance, in anticipating the pumping crisis that The Big One might initiate, it turns to a 1947 hurricane to describe water levels, 1965's Hurricane Betsy to measure pumping capabilities, and the 1900 hurricane in Galveston, TX that literally washed the city away and prompted the island on which it was built to be raised seven feet. At various points, in describing everything from logistical response strategies to changes in disaster policy, the report cites numerous hurricanes, earthquakes, tsunamis, wildfires, and floods around the world to illustrate exactly the stakes of disaster in New Orleans. The last sentence of this first section paints a bleak apocalyptic scenario for New Orleans: "Without extraordinary measures, key ports, oil and gas production, one of the nation's most important fisheries, the unique bayou culture, the historic French Quarter and more are at risk of being swept away in a catastrophic hurricane or worn down by smaller ones."[54] The reference to "smaller ones" appears as more than lip service to the regularization of living with hurricanes in the city, however, because the report then immediately turns to the next section, "The Big One."

Fairly straightforward in its logic in the first—development has increased risk generally—the second installment of the report introduces the spectral presence that animates it throughout. Its subtitle signals the contradictory logic at the heart of The Big One: "It's a matter of when, not if. Eventually a major hurricane will hit New Orleans head on, instead of being just a close call. It's happened before and it'll happen again."[55] Below the subtitle is an image of the aftermath of Hurricane Georges (1998). In it, a child stands among the wreckage of what can only be presumed to have been a lakeside house. Angled roughly parallel to the coast, the perspective of the photo affords a view of a decimated waterline reduced to battered structures and debris. And yet, the report notes, Georges was a near miss that "measured the slender margin separating the city from mass destruction." The routine occurrence of major hurricanes in the Gulf Coast may be just another part of life on the Coast, but in New Orleans there looms a Big One, the one that jeopardizes the city itself, the one for which Pam owes its very existence, and "officials say that right now, nothing can stop 'the big one.'"[56] The indeterminacy between routine and singular marks this anticipation as anxiety: although "it's happened before," The Big One's singular tense defines its singularity and inevitability as an event. Not if, but when.

This section of "Washing Away" then turns to Joseph Suhayda, an oceanographer who developed computer simulations of the storm, to narrate a technical description of The Big One that rivals portions of the SLCH report. Suhayda describes a "worst case" scenario of a hurricane approaching "from due south of the city." With the help of Suhayda, the special report narrates the movement of the hurricane, the rising water levels, and the overtopping and perhaps even failure of the levees. "There would be no stopping or slowing it," Suhayda says, "pumping systems would be overwhelmed and submerged within a matter of hours." As the narration, too, breaches the levees and enters the city, the descriptions become even more vivid:

> As the floodwaters invade and submerge neighborhoods, the wind will be blowing at speeds of at least 155 mph at ground level, accompanied by shorter gusts of as much as 200 mph, meteorologists say, enough to overturn cars, uproot trees and toss people around like dollhouse toys. . . . Ninety percent of the structures in the city are likely to be destroyed . . . Amid this maelstrom, the estimated 200,000 or more people left behind in an evacuation will be struggling to survive.

After the storm, the water and food necessary for such survival will have been contaminated by both the storm and the wildlife also struggling for survival. These conditions will have become a breeding ground for disease and other illness. "The Big One" closes by citing a public health expert at the LSU School of Medicine in New Orleans: "History repeats itself. . . . My office overlooks one of the St. Louis cemeteries, where there are many graves of victims of yellow fever. Standing water in the subtropics is the breeding ground for mosquitoes."[57] Bookended by the tragedies of New Orleans's past, "The Big One" is an imagination of the city's destruction and, unlike Pam, relies directly on the sorts of failure that render it thinkable. The scientific discourses that lend authority to this imaginative destruction may produce the warrant for such thinkability; the culmination of the report, however, is the apotheotic visual rendering of this destruction that reflects a desire not simply to think it, but also to frame a warning around the desire to witness it.

It is a fascinating case of extrapolation, given the fairly delimited function of the computer models as described by the report. As it notes, the acceleration of computational capacity has allowed these simulations to be more complex, and to be run iteratively more times through more scenarios. Much like in the past, the authors note, "scientists use 'hindcasting' to make it conform to reality. They plug the parameters of a real storm into a model and compare the results with recorded storm-surge heights. If the numbers don't line up, then the program needs adjusting." Methodologically, "hindcasting" is no more than model verification or validation, fitting the model to the historical record to ensure accuracy. This remains a standard practice in meteorological research using the complex models cited in the report. Having completed this verification, modelers then run the scenario through as many variations as is practicable, changing parameters as needed. The outputs, however, are fairly strictly technical: they predict storm surge levels, levee overtopping, and other hydrological dynamics that may be relevant to designing against disaster conditions. The use of the term "reality" here, then, articulates what are essentially distinct phenomena: On the one hand, the hydrological conditions that models and simulations are designed to predict, and, on the other, the benchmarks of misery poeticized through the narrative of destruction rendered by the imaginative response to these models. It is an articulation already made by Suhayda in the perspectivation of the model, prior to the popularization of the research through the news report.

Finally, the penultimate section of the report, "Tempting Fate," uses numerous images of the destruction wrought by Hurricane Andrew to frame the central political argument of the entire series, that irresponsible development has created "a new era" of mass destruction that threatens human existence. In nearly every part of this section, details of the heightened risks that development has spurred are accompanied by requisite images of exploded houses, messy tangles of debris, downed trees, and entirely flattened neighborhoods. One image shows a helicopter-eye view of a suburban intersection, houses flanking each corner and lining the receding streets. While the houses remain standing, the roofs have been stripped to their skeletal wood framing. Amidst the dusting of general debris, cars sit at odd angles, ripped-out trees lie in yards and across roads, and there is not a human in frame. While the scope of this section widens to include megadisasters across the nation, its logic rests firmly on the anxiety that Morris argues characterized New Orleans from the beginning: "As people have tried to tame nature by building homes, redirecting water, suppressing fires and reshaping coastlines, they have disrupted or blocked natural processes. But you can't just lock nature in place, and these measures have accelerated cycles of destruction in unpredictable and dangerous ways." Development, that *pharmakon* of the always latent disaster of civilization itself, progresses immediately to an overextension felt only as such in those moments when nature "struck back." Flanked by images of actual ruin, these warnings not only make the case for "not if, but when"; they also prefigure the landscape of New Orleans in the aftermath of The Big One. That is to say, they organize a visual field of ruin that functions not only as an "archaeology of the present," as Hariman and Lucaites detail it, but also as the very frame of reference for the post-disaster rememberer. Alongside more traditional sites of memorialization, ruins and destruction emerge as a primary object of memory, drawing thousands of tourists to the city to experience the disaster themselves, already conditioned as to not just what to see, but also how to see.

POST-KATRINA'S MEMORY OBJECTS

The endlessly repeated narratives of The Big One, as well as the retrospective assessments of why those narratives failed to prevent what was known to be imminent, must be read as iterative performances of an anxious pre-mourning process. The Big One was not merely the hurricane itself, which was anticipated and predicted with horrifying accuracy, but rather as an anxious rhetoric, it enters the realm of what Donald Rumsfeld famously called the "known unknowns," the things we know that we do not know.[58] While Slavoj Žižek has argued persuasively that disbelief in the possibility of ecological catastrophe has largely guided the lack of inaction to prepare for it (for him, "unknown knowns," or unconscious disavowal),[59] this has never been the case in narratives of The Big One. If "unknown knowns" represent disavowal, "known unknowns" can be serviceably translated as anxiety. I have argued that this anxiety and the discourse that marks it prefigured Katrina's memory. On one level, they create the conditions whereby the HSGAC hearing can retroactively name Katrina as The Big One in the first place. On another, however, they prefigure the very rhetorical form of melancholic mourning within the specific object domain within which public memory studies normally operates.

Critics and scholars studying New Orleans have long addressed the question of memory in the aftermath. At regular intervals in its 300-year history, major catastrophes have left in their wakes the question of abandonment: to rebuild the city or not? As Ari Kelman summarizes it, "[s]hould a place like New Orleans, in so dangerous a location, be reconstructed, especially given the costs and the likelihood (or near-certainty) that the next disaster is not a matter of if but when?"[60] Like Morris, Kelman recognizes disaster anxiety as part of the city's narrative. However, he ups the historical ante by reading it as not only a product of an originary tension between nature and development, but also as the necessary capacity to forget in order to recover:

> New Orleans has a horrible disaster history. The city, it might seem, hardly finishes counting bodies before the next tragedy strikes. Katrina, I think, will become just one more chapter in that long saga. Epidemics, fires, floods, and hurricanes: after each visitation, the city has picked itself up and rebuilt, largely because of an uncanny ability to forget past tragedies and ignore the next one

> lurking around the corner. New Orleans changes after grappling with disaster, to be sure. But it's a resilient place, practiced in the art of recovery and forgetting.[61]

Kelman's book serves as a testament to this claim, though perhaps in more ways than he intends. Originally published in 2003, the book details the history of the riverfront in New Orleans through a socioeconomic historical lens, demonstrating the various ways in which manipulation of and control over the river's banks have shaped the relationships between the cityscape, nature, capital, class, and public space. Traced across each era of its history, New Orleans is understood as a center of capitalistic opportunity whose existence is not only a product of its ability to reach a forceful equilibrium with "nature," as Morris argues, but also its reliance on a calculus that determines whether in any given case that equilibrium will be a profitable one. Read through such a lens, the city's resilience is framed less as an art of forgetting and more as a matter of economic advantage.

However, the above quote appears in the preface to the paperback edition, published in 2006. In it, Kelman recalls lying in bed the night of the storm, "thinking that the city had dodged another bullet. This storm, though massive and powerful, wasn't going to be New Orleans's 'big one,' the killer hurricane that some day would make its way directly up the Mississippi River or over Lake Pontchartrain, dumping a massive storm surge on the city."[62] "Some day," as I have said before, rather than "someday": not some indeterminate time in the future, but a single, apocalyptic day to come. This would lead the reader to believe that, after the levees broke, after images of the drowned began circulating, after the horrific (though often false) stories of conditions inside the Superdome spread, and even after the administration acknowledged its failure to respond adequately, Katrina would indeed have been The Big One foretold by popular, institutional, and even academic mythos. Yet, Kelman's conclusion is that, in the final analysis, the memory of Katrina will be both a remembering and a forgetting, that, figured similarly to the images of Hurricane Georges's devastation that represented a "near miss," Katrina will be slotted into the long list of catastrophes that merely precede the next inevitable tragedy.

Michael Bibler's examination of *Jezebel*, a 1938 film about the aforementioned 1853 yellow fever epidemic in New Orleans, demonstrates exactly this point. Rewatching the film post-Katrina, he recounts being "surprised and frightened" at the parallels between the film's scenes and the 24-hour news coverage of Katrina: "As I watched the film wind to its conclusion, I began to realize: the film wasn't depicting something analogous to Katrina as much as it was almost *predicting* Katrina."[63] The film's depictions of the racial and economic factors at work in the ability to evacuate the city, the quarantining of the city and the official orders to shoot anyone trying to escape, the fumbling actions of city officials to prevent and respond to the epidemic; these and other scenes appear to do more than simply fantasize about how a major natural disaster scenario would play out in New Orleans. They appear to predict it—*almost*. Rather than hastily giving into this conclusion, Bibler articulates it to the fantasy of a "perpetually doomed southern city" and instead argues that "we can view much of the discourse from Katrina's aftermath as having been shaped by a centuries-long American perception that New Orleans is destined for a tragic ending—a myth that the city itself has done little to dispel."[64] The film's title, *Jezebel*, sheds additional light on that "uneasy tension" that symptomatizes Morris's characterization of the region. Jezebel: "the city that is always headed toward destruction because of its decadence, its anti-modernity, and for many, its blackness."[65] These two academic texts, one retrospective and one critical, provide an inroad for tracing the incorporation of anxious discourse into the memory objects of a post-disaster exigence. This troubles the commonly understood relationship between remembering and forgetting in the context of a perpetual repetition of disaster—in its occurrence and in its figuration.

The Hurricane Katrina Memorial

Consider first the Hurricane Katrina Memorial in New Orleans. During a ceremony two years after Hurricane Katrina made landfall, then-Mayor C. Ray Nagin announced that $1.2 million would be allocated towards building a proper site of remembrance for the victims. This event was billed perhaps prematurely as "breaking ground on a permanent memorial that [would] also contain the remnants of several dozen unclaimed bodies."[66] Nagin's ceremonial address was delivered inside Charity Hospital Cemetery, nestled between Cypress Grove Cemetery and St. Patrick Cemeteries Nos. 1 and 2 and soon to be known as Charity Hospital & Katrina Memorial Cemetery or, simply, the Hurricane Katrina Memorial. The site is historically known as "a last resting place for unclaimed dead."[67] Formally dubbed "Potter's Field" at the time of its establishment in 1848, the biblically christened site was deemed a suitable resting place for unidentifiable bodies, and later included those who had died of rampant contagious diseases. "In the 1800s," writes John Salvaggio, "all unclaimed bodies of the dead in New Orleans were buried there, but by the mid-1900s it was used only for the unclaimed bodies of patients who had died in Charity Hospital."[68] The city removed many of those remains during the construction of the memorial. Charity Hospital Cemetery appears, at first glance, a most fitting location for commemorating these victims of Katrina: it is a place whose historical narrative renders it singularly capable of memorializing the unattached.

However, it was not long before the Hurricane Katrina Memorial appeared to be forgotten altogether. Funding was not at issue: the money had been secured through the city and private donors, and long-term maintenance costs had also been allocated.[69] Nor was there any heated public debate over design, location, or the question of how to name the victims. In the end, because "a group of funeral home owners . . . took it upon themselves to inter the remains because they felt the city and coroner's office were too slow to do so," the city raced frantically to finish construction prior to what would be an inspired third-year anniversary ceremony.[70]

This timeline stands in stark contrast to other memorialization practices for major national tragedies. In the early 1980s, for instance, the Vietnam Veterans Memorial in Washington, D.C. became a subject of national controversy. Because of its unconventional design, a black granite wall cut into the earth and engraved with the names of fallen soldiers, the public outcry was so great that the federal Commission of Fine Arts added a large flagpole and a sculpture of three soldiers. While the artist of the original design, Maya Lin, imagined it as a cut into the earth, "an initial violence and pain that in time would heal,"[71] detractors thought the wall looked like a trench and was ultimately a "symbol of shame or dishonor for veterans."[72] The addition of the flagpole and statue, they argued, more properly valorized the sacrifice of the fallen soldiers. As Carole Blair, Marsha Jeppeson, and Enrico Pucci, Jr. argue, the debate over the memorial was nothing less than a political debate over the meaning of the Vietnam War itself, the final combination of memorial objects that "inscribe a history, forming a space of cooperative conflict and commenting on each other's statement about the war."[73] Likewise, the design for the National September 11 Memorial & Museum "needed to take into account the various opinions and agendas of more than 30 consulting parties, including government agencies, preservationists, and groups that represented the surrounding community and survivors and victims of the attacks."[74] Moreover, the decision to sell merchandise and charge entrance fees to the museum,[75] the proposal to list the names of the victims level with the ground,[76] and the omission of the history of Little Syria that formerly occupied the space where the Twin Towers later stood[77] were all topics of public debate. Among the more heated topics was the plan to inter the unidentified remains of 9/11 victims underneath the museum. One victim's family member, arguing on behalf of protesters at the ceremony for the remains' return to the site, said, "We would like to see remains moved above ground in a repository akin to the tomb of the unknowns. . . . On the plaza, to add a little respect and dignity."[78] While the loss of life in both the Vietnam War and 9/11 outstrips Katrina, official estimates put Katrina's death toll at 1,833 and its cost at $108 billion.[79] Katrina was a staggering national tragedy from which President Bush's presidency would never quite recover. Yet, the nation saw almost no public deliberation about how best to memorialize it, either in the articulation of the tragedy's public meaning or in its specific treatment of the unclaimed and unknown dead. Even after the memorial was constructed, just prior to the ten-year anniversary of the storm, ABC's local New Orleans news station published a spotlight on "The Katrina Memorial You Don't Know About," that characterizes it as a "place where the dead seem to have been forgotten."

On this level, the memorial painfully reenacts the systemic, violent exclusions that led to many of Katrina's gravest social consequences in two ways. First, it capitalizes on the synecdochal premise of the Tomb of the Unknown Soldier at Arlington, but is itself located within a resolutely ambivalent memorial space and performs that ambiguity in the slippage between "unclaimed" and "unknown." As Blair, Balthrop, and Michel have deftly demonstrated, quite often "unknown" is a signifier produced explicitly to comport with rhetorical appeals that require such anonymity to function in public memorialization, and the Katrina Memorial is no exception in this regard. They detail the numerous ways in which nations have sought to ensure that soldiers buried within various memorial sites commemorating unknown war casualties remained anonymous, including removing all potential identifying information, destroying original burial records, and even assigning blind or blindfolded veterans to select the remains.[80] Securing the anonymity of the remains, they contend, enabled the enthymematic premise forwarded by the memorials' inscriptions: "*This could be your husband, your father, your brother, your son* . . . "[81] While less concerted in method, the signification of unknown status of the bodies interred at the Katrina Memorial was not precisely true. To the contrary, Frank Minyard, longtime Orleans Parish coroner and conceiver of the memorial, actually knew some of the names; he only lacked permission to use them.[82] Even more, there was concern that officially naming these dead as victims of the hurricane (and not of the subsequent flooding, exposure, lack of medical access, and/or the acute effects of stress) would categorically compromise the insurance compensation of victims of the storm's immediate and long term effects, up to and including life insurance policies.[83]

Nevertheless, from speeches to inscriptions, the unclaimed and the unknown are made coextensive through their conflation, reperforming the anxiety of disaster (the known unknown) by producing unknownness as exception. As a veritable Tomb of the Unknown Hurricane Victim, the city claims these bodies as its own, immortal in memory because singular in function, though not in name.[84] Take, for instance, an excerpt from then-Mayor Mitch Landrieu's speech delivered throughout the ninth-year anniversary ceremony as city and community leaders finally consecrated the entombment of these unclaimed dead:

> They may be unnamed, but they are not unclaimed. We claim them, as ours. Let us make them proud, because *that* is how they can rest in the peace they have so utterly deserved. And the unnamed men and women in this sacred space are gonna help lead us to a New Orleans that should have always been.[85]

Though the names of the bodies are (sometimes deliberately) forgotten, the bodies themselves are nevertheless remembered. Remembrance of the victims presumes to bring the city closer to its ideal—except in practice, these bodies are once more produced as unnamed and relegated to a site of forgetting (a potter's field) historically designed just for them. As Lindsay Tuggle notes, "[t]he Katrina Memorial contains bodies that interrupt dominant cultural narratives of patriotism and privilege," bodies which are also "consigned to a specified, bordered location on the outskirts of town."[86] Understanding the remains in this way comports with the general consensus—both scholarly and public—regarding Katrina's memory writ large: political economic systems, from economic and racial segregation to infrastructural failures, produced a disaster far more destructive and insidious than any single storm could have. As Michael Dyson notes regarding Katrina, the country was "immediately confronted with [an] unsavory truth: it is the *exposure* of the extremes, not their existence, that stumps our sense of decency."[87] In the conflation of unknown and unclaimed, the Katrina memorial incorporates the anxiety of the known unknown, the social disaster the storm exposed and exploited, into its very architecture.

Second, this rhetorical architecture also demonstrates the coextensivity of the city and its hurricane in its physical metaphorization of a hurricane. A satellite image of the site (Figure 4.1) reveals its hurricane-shaped construction: concrete sidewalks spiral outward from the center, lined on their outside edges with the mausolea interring the dead, on the other side of which is planted two symmetrical arcs of trees that engulf the site completely. At the center of the eye of the storm is mounted a black granite stone, which explains:

> Most of the deceased were identified and buried by loved ones in private ceremonies throughout the nation. Here lie the remaining. The unclaimed and unidentified victims of the storm from the New Orleans area. Some have been forgotten, some remain unknown.
>
> This memorial is dedicated to these individuals and to all who suffered or died during Hurricane Katrina. Let the victims here forever remind us of those harrowing days and the long struggle to rebuild our city. Let their final resting place call us to constant preparedness. Let their souls join into an eternal chorus, singing with the full might of the indomitable spirit of New Orleans.

Sheltered within the epideictic spirit of this inscription is the deliberative appeal to "constant preparedness," a reminder that The Big One is merely an anticipation that is, to channel Derrida once more, "*fabulously textual*, through and through."[88] While the outward purpose of the site is to honor the dead, the status of the dead within it is a reminder that, still, it is not a matter of if, but when.

The inner rain bands, made partly of the unknown and forgotten dead, metaphorize precisely the environmental tension that has always characterized New Orleans: the city on the inside, its "indomitable spirit" placed directly in the center, and nature on the outside, trees that lie just outside the concrete partitions that institute this division. Acting as the levee between the concrete inside and the natural outside, the six mausoleums, nameless and forgotten casualties pressed between concrete and earth: they become the technology that failed them. Between the dry and the wet, the forgotten and the unknown are not only conflated in speech and inscription, but also are they combined physically in this space, quite literally incorporated—*incorporāre*: to "combine or form into one body"[89]—into its construction. There is perhaps no product of anxious melancholic rhetoric more apt than the co-entombment of the forgotten and the known unknown. In this capacity, the memorial positions them carefully within a process of mourning already anticipated, already sublimated. Like Hurricane Pam's levees, the forgotten/unknown are the contingencies of failure, exceptional in the fullest sense, and are thus emblematic of the indomitable spirit of New Orleans.[90]

Figure 4.1. Satellite view of the Hurricane Katrina Memorial in New Orleans. *Source*: Imagery © 2018, United States Geological Survey.

The Bus Tours

Another popular memory practice that emerged after the storm was the organization of bus tours that surveyed affected areas. Buses filled with tourists regularly drove the streets of New Orleans that were the hardest hit by the hurricane and flooding. "In fact," writes Lynnell Thomas, "by 2007 just about every tour company, across a broad spectrum of tourism genres, was offering some type of Hurricane Katrina-related tour."[91] The tours became immediately controversial as busloads of what Phaedra Pezzullo would recognize as a version of "toxic tourists"[92] rolled into ravaged neighborhoods—the residents of which were often still actively cleaning up their properties—snapped photos, and moved on. Pezzullo contends that, as memory practices, these tours are forms of witnessing that can be powerful counters to "disaster fatigue." To that end, "[w]hether or not non-residents of the region have become too tired to recall, or pay attention, is articulated to whether or not the nation will be willing and able to rebuild and to take collective responsibility."[93] This is no doubt true to some degree, and the present aim is not to deny such political possibility. Indeed, from the beginning, many of these tours have been guided by long-term residents who lost their homes during Katrina and the tours themselves have always donated a portion of the cost to rebuilding efforts.[94] Nevertheless, these tours are borne not only of a New Orleans tourism industry marked by both desire and disaster, but also of a visual field of ruin that marks the discourses of The Big One.

As a logically fitting outgrowth of the New Orleans tourism industry, the bus tours put disaster into the service of memory by remapping part of the tourism landscape and by instituting a physical barrier between inside and outside, seen also in the Katrina Memorial—but this time, coded more explicitly as racial difference articulated through what Anna Hartnell critiques as a "satisfying blend of proximity and distance."[95] Lynnell Thomas notes that "whereas pre-Katrina city tours strategically circumnavigated the city to avoid most historically and predominantly African American neighborhoods, post-Katrina tours consistently remapped tourist New Orleans to include African American spaces in the Upper and Lower Ninth Wards, parts of New Orleans East and Gentilly, and the Tremé."[96] That is, whereas before The Big One, the racial poverty and inequality that have always undergirded disaster outcomes and policy in the South[97] remained out of the frame of reference in memory, in post-Katrina memory practices they enter the frame directly. As Greg Hoffman, vice president of Gray Line Tours and home-losing resident of New Orleans put it, "Well, we know that a lot of visitors are going to want to go to ground zero and out to the devastated areas, so we're trying to give them an option in a safe and organized fashion."[98] "Ground Zero," as a post-9/11 metaphor insofar as the event ripped the metaphor from the context of the atom bomb, was invoked consistently across the news mediascape, drawing on both tragedy and destruction, but also orienting these sites of destruction as the primary places of public memory where one would come most fully to appreciate and to mourn the trauma that Katrina represented. As another guide for a different tour company told reporters, "I want to show people plantation homes but they don't want to see it. . . . They want to see the disaster."[99]

This "seeing" occupies an ambivalent political, visual space that oscillates between voyeurism and witnessing, but in either capacity establishes a distance that is a prerequisite for establishing the visual field of ruin that "Washing Away" assists in prefiguring. A central debate in both the popular and academic domains questions the normative valences of each of these modes of seeing. News media, in their ever reliable fetishization of political binarism, reported consistently that some residents believed that exposing visitors to the destruction firsthand would engender greater political will to push for assistance, while others found the practice disrespectful to those who had lost everything. Likewise, while Pezzullo, for instance, sees post-Katrina tourism as opening spaces of advocacy, other scholars like Kevin Fox Gotham argue that the "bus tour present flooded neighborhoods as spectacular and entertaining sites to visit."[100] Whatever the ethics of this practice at the level of individual seeing, which is most positively a philosophical and political determination, the rhetorical function of the tours remains to institute that longstanding division between inside and outside, in New Orleans, always coded as the wet and the dry, the city and nature. As Hoffman explains, for instance, "[p]eople are not allowed to get off the bus, to take any pictures."[101] Another news report repeats the line: "Pictures are permitted, but no one is going to be allowed to get off the bus and walk around."[102] In both instances, the acknowledgement that visitors would not be allowed off the bus—undeniably a good thing—is nevertheless intelligible only alongside the concomitant assurance that pictures would still be allowed. Visitors, more likely after Katrina to desire to gaze upon the ruin of the city's most disadvantaged, document the city's ruin, producing their own photos striking in similarity to the ones that "Washing Away" includes in its imaginative warnings. Afterwards, they "are safely deposited back in the French Quarter,"[103] disaster archive in hand. The tours enact a mnemic voyeurism, a documentation of the aftermath of The Big One that reproduces the images that anticipate it.

The tours do more than simply accommodate the desire to view disaster from a safe distance, however: in staging a field of ruin largely devoid of human life, they displace the social costs of this devastation in favor of the environmental narrative of development that was the premise of "Washing Away." Around the time the tours started, New Orleans's population remained at roughly 35 percent what is was prior to the storm, and the highest concentration of those displaced or permanently relocated people, and by extension those most likely to have houses destroyed by the storm and flooding, once resided in the very spaces that these tours traversed.[104] Of course, the tours regularly encountered residents cleaning up, who received the visitors with the same mixed attitudes as news and academic sources relate. By and large, however, as Hartwell suggests, "where the media coverage of the storm fixed its eye on human beings, Katrina tourism takes in scenes that are, on the whole, peculiarly devoid of human inhabitants."[105] Like the images of post-Andrew Florida and the post-Georges Gulf Coast, the visual field that Katrina bus tours provide is one of mass ruin, an archaeology of the present that signals a post-human landscape of environmental devastation.

This, precisely, is the incorporation at work in the bus tours. While the very touristic geography of the city was remapped and reorganized around the devastation of its most vulnerable spaces, the visual field that the tour provides makes up a landscape largely devoid of humans and the social. The narrative of development that structures "Washing Away" and its imaginary hurricane signals a lesson and a warning, not of the social and economic sort, but of the environmental sort. The presumed failures on which it relies are primarily infrastructural, not social: the levees will breach, the pumps will fail, and the city will fall to ruins. Likewise, Katrina bus tourism does not so much tell the story of "the environment's inextricable links to social justice issues—such as safe living spaces, or the impact of climate change on the poor," argues Hartwell, "but rather . . . creates a consoling narrative around the preservation of a shared natural habitat."[106] By largely ejecting the presence of the human from the visual frame, they, like "Washing Away," instead incorporate the image of post-human ruin that displaces the social disparities that produce such devastation in the first place. In so doing, these memory objects organize the memory of Katrina around a general conception of the hubris of overdevelopment and the wrath of nature's response—all within a city that has a long and storied history of enduring both, but all without the direct confrontation of the racial and economic traumas that Katrina most centrally exposed. The visual field of The Big One activates a generic tale of the absolute failure of environmental balance wherein such acknowledgement is not required.

CONCLUSION

It is only after Katrina that Kelman accedes to the melancholic premise of "The Big One." Whereas in his original analysis, development in New Orleans progresses *despite* a clear knowledge that the city was particularly prone to disaster, after Katrina he reformulates this thesis to claim that development in New Orleans progresses *because of* its environmental and social vulnerability, "*because of* an uncanny ability to forget past tragedies and ignore the next one lurking around the corner."[107] Neither "forget" nor "ignore" are perhaps quite the right words, however, as I hope to have shown in this chapter. Indeed, the dialectic between remembering and forgetting that animates public memory is what makes possible the notion that memory practices in New Orleans can incorporate the anxieties of potential failures, all the while over and over again anxiously forecasting its own demise.

As further confirmation, *The Times-Picayune* published yet another special report entitled "Our Drowning Coast" in February 2018. Focusing mainly on environmental factors rather than strictly developmental ones, the parallels between it and "Washing Away" are striking. "This year, New Orleans celebrates its 300th birthday," it informs. "Whether it will see 400 is no sure thing." As the report summarizes the post-Katrina infrastructure that federal, state, and local governments spent more than $20 billion on, it notes that these protections are probably not nearly enough, given the environment surrounding the city and the stakes of another Big One. The Army Corps of Engineers, in fact, no longer considers this system a "hurricane protection system," opting instead to call it a "hurricane and storm damage risk reduction system." The author ruminates, "How this came to be is a story of money and politics and, perhaps, a degree of Louisiana fatalism."[108]

While on more celebratory days, tourists mill up and down Bourbon Street with Hurricanes in hand (a cocktail of rum, fruit juice, and grenadine) and locals taunt imminent storms with hurricane parties, in more anxious ones, they tell the story of their own city's demise. Given the melancholic sequence of these narratives, it is impossible to define public memory of Katrina outside of them and the ways in which they are incorporated into its memory objects. Far from resolving the anxieties that such stories produced prior to Katrina or the loss felt after, these objects incorporate failure in their rhetorical renderings, and thus employing rhetoric as a mode of containment.

NOTES

1. A version of this chapter was previously published as: Jeremy R. Grossman, "The Catastrophe to Come: Prefiguring Hurricane Katrina's Public Memory through the Anxious Melancholic Rhetoric of 'The Big One,'" *Rhetoric & Public Affairs* 23, no. 3 (2020), 417–53.

2. John McQuaid and Mark Schleifstein, "Special Report: Washing Away," *The Times-Picayune*, June 23–27, 2002.

3. Innovative Emergency Management, *Southeast Louisiana Catastrophic Hurricane Functional Plan*, August 6, 2004, 7.

4. Innovative Emergency Management, *Southeast Louisiana*, 7.
5. Innovative Emergency Management, *Southeast Louisiana*, 18.
6. Innovative Emergency Management, *Southeast Louisiana*, 27.
7. Innovative Emergency Management, *Southeast Louisiana*, 31.
8. Innovative Emergency Management, *Southeast Louisiana*, 43.
9. Innovative Emergency Management, *Southeast Louisiana*, 48.
10. Innovative Emergency Management, *Southeast Louisiana*, 59.
11. Innovative Emergency Management, *Southeast Louisiana*, 76.
12. Innovative Emergency Management, *Southeast Louisiana*, 92.
13. Innovative Emergency Management, *Southeast Louisiana*, 104–5.
14. Cedric Johnson, "Introduction," in *The Neoliberal Deluge: Hurricane Katrina, Late Capitalism, and the Remaking of New Orleans*, ed. Cedric Johnson (Minneapolis: University of Minnesota Press, 2011), xxxi.
15. Madhu Beriwal, *Preparing for a Catastrophe: The Hurricane Pam Exercise*, statement before the Senate Homeland Security and Governmental Affairs Committee, January 24, 2006, *https://www.hsgac.senate.gov/imo/media/doc/012406Beriwal.pdf* (accessed September 26, 2018).
16. "Back to the Future in the San Andreas Fault: Investigating Past Earthquakes to Inform the Future," *U.S. Geological Survey*, https://earthquake.usgs.gov/learn/topics/safz-paleo/ (accessed September 26, 2018).
17. Kathryn Schulz, "The Really Big One: An Earthquake Will Destroy a Sizeable Portion of the Coastal Northwest. The Question Is When," *The New Yorker*, July 20, 2015, https://www.newyorker.com/magazine/2015/07/20/the-really-big-one (accessed January 5, 2024).
18. Hugh Riminton, "Indonesia's Big One 'On Its Way,'" *CNN*, September 18, 2007, http://edition.cnn.com/2007/WORLD/asiapcf/09/17/indonesia.quake (accessed September 26, 2018).
19. Natalie Wolchover, "The Big One: When Will the Yellowstone Supervolcano Erupt?" *Live Science*, June 8, 2011, https://www.livescience.com/33330-yellowstone-caldera-supervolcano-eruption.html (accessed September 26, 2018).
20. Barbara A. Biesecker, "No Time for Mourning: The Rhetorical Production of the Melancholic Citizen-Subject in the War on Terror," *Philosophy and Rhetoric* 40 (2007): 154.
21. Carole Blair, Greg Dickinson, and Brian L. Ott, "Introduction: Rhetoric/Memory/Place," in *Places of Public Memory: The Rhetoric of Museums and Memorials*, eds. Greg Dickinson, Carole Blair, and Brian L. Ott (Tuscaloosa: University of Alabama Press, 2010), 18.
22. Vivian, *Public Forgetting*.
23. Blair, Dickinson, and Ott, "Introduction," 18.
24. Huet, *The Culture of Disaster*, 7.
25. Lynnell Thomas, *Desire and Disaster in New Orleans: Tourism, Race, and Historical Memory* (Durham, NC: Duke University Press, 2014).
26. Morris, *The Big Muddy*, 1.
27. Jacques Derrida, *Dissemination* (Chicago: University of Chicago Press, 1981), 61–171.

28. Morris, *The Big Muddy*, 219.

29. Sigmund Freud, *Introductory Lectures on Psycho-Analysis*, trans. James Strachey (New York: W. W. Norton, 1966), 510.

30. Freud, *Introductory Lectures*, 507; in the original source, he refers to a "damming-up" of libido.

31. Sara Beardsworth, *Julia Kristeva: Psychoanalysis and Modernity* (Albany: State University of New York Press, 2004), 196.

32. Beardsworth, *Julia Kristeva*, 197.

33. Robert Hariman and John Louis Lucaites, *The Public Image: Photography and Civic Spectatorship* (Chicago: University of Chicago Press, 2016), 121.

34. Hariman and Lucaites, *The Public Image*, 127.

35. For an elaboration of this in the context of melancholy and subjectivation *vis à vis* rhetoric, see Biesecker, "No Time for Mourning," 148–52.

36. Judith Butler, *The Psychic Life of Power: Theories in Subjection* (Stanford, CA: Stanford University Press, 1997), 132.

37. Butler, *The Psychic Life of Power*, 134, emphasis in original.

38. Butler, *The Psychic Life of Power*, 134, emphasis in original.

39. Sigmund Freud, *The Ego and the Id*, trans. James Strachey (New York: W. W. Norton, 1960), 24.

40. U.S. Congress, Senate, Committee on Homeland Security and Governmental Affairs, *Preparing for a Catastrophe: The Hurricane Pam Exercise*, 109th Cong., 2nd sess., January 24, 2006, 2.

41. *Preparing for Catastrophe*, 11.

42. *Preparing for Catastrophe*, 16.

43. FEMA, "Hurricane Pam Exercise Concludes," July 23, 2004, https://biotech.law.lsu.edu/blaw/FEMA/FEMA_Hurricane_Pam.pdf.

44. Jacques Derrida, "No Apocalypse, Not Now (Full Speed Ahead, Seven Missiles, Seven Missives)," *Diacritics* 14 (1984): 24.

45. *Preparing for Catastrophe*, 2.

46. *Preparing for Catastrophe*, 3.

47. Brian Handwerk, "New Orleans Levees Not Built for Worst Case Events," *National Geographic News*, September 2, 2005.

48. Innovative Emergency Management, *Southeast Louisiana*, 1.

49. Slavoj Žižek, *The Parallax View* (Cambridge, MA: The MIT Press, 2009), 30.

50. *Preparing for Catastrophe*, 8, emphasis added.

51. This can also be seen in a very similar 1959 model called the "standard project hurricane" developed by the Weather Bureau and the Army Corps of Engineers, wherein, in calculating the features of the hypothetical storm, "[s]torms like Hurricane Camille in 1969 were taken out of the data set as lying too far out of the norm. . . . Also, the calculations of the cost-benefit ratio did not take into account the costs of failure, both economic and social, far greater in an urban area like New Orleans than a rural one." John Schwartz, "An Autopsy of Katrina: Four Storms, Not Just One," *The New York Times*, May 30, 2006, *https://www.nytimes.com/2006/05/30/science/30storm.html* (accessed January 5, 2024).

52. *Preparing for Catastrophe*, 14.

53. John McQuaid, "Washing Away: In Harm's Way," *The Times-Picayune*, June 23, 2002.

54. McQuiad, "In Harm's Way."

55. John McQuaid, "Washing Away: The Big One," *The Times-Picayune*, June 24, 2002.

56. McQuaid, "The Big One."

57. McQuaid, "The Big One."

58. Donald Rumsfeld, "Department of Defense News Briefing," *U.S. Department of Defense*, February 12, 2002, *https://usinfo.org/wf-archive/2002/020212/epf202.htm* (accessed January 5, 2024).

59. Slavoj Žižek, "Joe Public v the Volcano," *New Statesman*, April 29, 2010, *https://zizek.uk/2010/04/29/joe-public-v-the-volcano/* (accessed January 5, 2024).

60. Ari Kelman, *A River and Its City: The Nature of Landscape in New Orleans* (Berkeley: University of California Press, 2006), xiii.

61. Kelman, *A River and Its City*, xii.

62. Kelman, *A River and Its City*, ix.

63. Michael Bibler, "Always the Tragic Jezebel: New Orleans, Katrina, and the Layered Discourse of a Doomed Southern City," *Southern Cultures* 14, no. 2 (2008): 7, emphasis in original.

64. Bibler, "Always the Tragic Jezebel," 8.

65. Bibler, "Always the Tragic Jezebel," 25.

66. Adam Nossiter, "Commemorations for a City 2 Years After Storm," *The New York Times*, August 30, 2007, sec. National / U.S., http://www.nytimes.com/2007/08/30/us/nationalspecial/30katrina.html.

67. Doug MacCash, "New Orleans Katrina Memorial Is Almost Perfect," *NOLA.com*, September 1, 2007.

68. John E. Salvaggio, *New Orleans' Charity Hospital: A Story of Physicians, Politics, and Poverty* (Baton Rouge: Louisiana State University Press, 1992), 50.

69. CBSNews/AP, "Katrina Victims."

70. CBSNews/AP, "Katrina Victims."

71. Maya Lin, "Making the Memorial," *The New York Review of Books*, November 2, 2000, http://www.nybooks.com/articles/2000/11/02/making-the-memorial/.

72. Carole Blair, Marsha S. Jeppeson, and Enrico Pucci, Jr., "Public Memorializing in Postmodernity: The Vietnam Veterans Memorial as Prototype," *Quarterly Journal of Speech* 77 (1991): 276.

73. Blair, Jeppeson, and Pucci, Jr., "Public Memorializing," 276.

74. John Caulfield, "9/11 Museum Triumphs over Controversy," *Building Design & Construction*, April 9, 2015, https://www.bdcnetwork.com/911-museum-triumphs-over-controversy.

75. Gio Benitez, "Why Victims' Families Are Furious about 9/11 Memorial Museum," *ABC News*, May 19, 2014, http://abcnews.go.com/US/victims-families-furious-911-memorial-museum/story?id=23774869.

76. David Dunlap, "Display of Names at Trade Center Memorial is a Painstaking Process," *The New York Times*, March 23, 2009, http://www.nytimes.com/2009/03/24/nyregion/24names.html.

77. David Dunlap, "Little Syria (Now Tiny Syria) Finds New Advocates," The New York Times, January 1, 2012, https://cityroom.blogs.nytimes.com/2012/01/01/little-syria-now-tiny-syria-finds-new-advocates/.

78. Joseph Stepansky, Erik Badia, and Larry McShane, "Some Family Members Denounce Return of 9/11 Remains to World Trade Center, Others Call It 'Disrespectful,'" *New York Daily News*, May 11, 2014, http://www.nydailynews.com/new-york/unidentified-remains-9-11-transfered-new-york-world-trade-center-article-1.1787160.

79. "Hurricane Katrina Statistics Fast Facts," *CNN*, August 28, 2017, https://www.cnn.com/2013/08/23/us/hurricane-katrina-statistics-fast-facts/index.html.

80. Carole Blair, V. William Balthrop, and Neil Michel, "The Arguments of the Tombs of the Unknown: Relationality and National Legitimation," *Argumentation* 25 (2011): 453–54.

81. Blair, Balthrop, and Michel, "The Arguments," 457.

82. MacCash, "New Orleans Katrina Memorial Is Almost Perfect."

83. CBSNews/AP, August 28, 2008, "Katrina Victims Are Buried, 3 Years After," accessed February 17, 2015, http://www.cbsnews.com/news/katrina-victims-are-buried-3-years-after/.

84. The Tomb of the Unknown Soldier has, in fact, been invoked in news reporting: MacCash, "New Orleans Katrina Memorial Is Almost Perfect."

85. "Hurricane Katrina Wreath-Laying Ceremony," *WDSU*, accessed May 12, 2019, http://www.wdsu.com/news/local-news/new-orleans/hurricane-katrina-wreathlaying-ceremony/27801328.

86. Linsday Tuggle, "Encrypting Katrina: Traumatic Inscription and the Architecture of Amnesia," in "The Cultural Visualization of Hurricane Katrina," eds. Nicola Mann and Victoria Pass, Special issue, *Invisible Culture* 16 (2011), http://www.rochester.edu/in_visible_culture/Issue_16/contributors.html#tuggle.

87. Michael Eric Dyson, *Come Hell or High Water: Hurricane Katrina and the Color of Disaster* (New York: Basic Books, 2006), 2–3, emphasis added.

88. Derrida, "No Apocalypse," 23, emphasis in original.

89. *The Oxford Dictionary of English Etymology*, s.v. "incorporate."

90. Lindsay Tuggle argues that the Katrina Memorial is, rather, an instance of "*failed* incorporation, establishing a site of perpetual detachment in which the other is permanently encrypted *without*, that is, *outside* of, the subject." It should be clear by now that this chapter seeks to read such incorporation not as the externalization of internal processes, but rather as a general function of melancholic discourse that threatens that very artificial distinction. Tuggle, "Encrypting Katrina."

91. Thomas, *Desire and Disaster*, 131.

92. Phaedra C. Pezzulo, *Toxic Tourism: Rhetorics of Pollution, Travel, and Environmental Justice* (Tuscaloosa: University of Alabama Press, 2009).

93. Phaedra C. Pezzullo, "Tourists and/as Disasters: Rebuilding, Remembering, and Responsibility in New Orleans," *Tourist Studies* 9 (2010): 26; see also Phaedra C. Pezzullo, "This Is the Only Tour That Sells": Tourism, Disaster, and National Identity in New Orleans," *Journal of Tourism and Cultural Change* 7 (2009): 99–114.

94. See, for instance, Greg Hoffman, interview by Soledad O'Brien, *CNN American Morning*, CNN, January 5, 2006.

95. Anna Hartnell, "Katrina Tourism and a Tale of Two Cities: Visualizing Race and Class in New Orleans," *American Quarterly* 61 (2009): 724.

96. Thomas, *Desire and Disaster*, 131.

97. Mark M. Smith, *Camille, 1969: Histories of a Hurricane* (Athens: University of Georgia Press, 2011), 21–35.

98. Greg Hoffman, interview by Neal Conan, *Talk of the Nation*, National Public Radio, December 13, 2005.

99. Erik German, "Next Stop: Katrina's Ruins; Bus Tours of Ravaged New Orleans Cause Controversy," *Newsday*, January 5, 2006.

100. Kevin Fox Gotham, "(Re)Branding the Big Easy: Tourism Rebuilding in Post-Katrina New Orleans," *Urban Affairs Review* 42 (2007): 838; see also Kevin Fox Gotham, "Touristic Disaster: Spectacle and Recovery in Post-Katrina New Orleans," *Geoforum* 86 (2017): 127–35.

101. Hoffman, *Talk of the Nation*.

102. Jeff Goldblatt, "Tour Group Offers Hurricane Katrina Devastation Tour," *The Big Story with John Gibson*, Fox News Network, December 26, 2005.

103. Hartnell, "Katrina Tourism," 723.

104. U.S. Library of Congress, Congressional Research Service, *Hurricane Katrina: Social-Demographic Characteristics of Impacted Areas*, by Thomas Gabe, Gene Falk, and Maggie McCarty, RL33141 (2005), 16–17.

105. Hartnell, "Katrina Tourism," 735.

106. Hartnell, "Katrina Tourism," 725.

107. Kelman, *A River and Its City*, xii, emphasis added.

108. John Schwartz and Mark Schleifstein, "Fortified but Still in Peril, New Orleans Braces for Its Future: Our Drowning Coast," *The Times-Picayune*, February 24, 2018, *https://www.nytimes.com/interactive/2018/02/24/us/new-orleans-flood-walls-hurricanes.html* (accessed January 5, 2024).

Chapter 5

Predictions of/and The Past

By now it should be abundantly clear that one of the central tensions at work in the science of prediction, conceptually and rhetorically speaking, is the relationship between the physical world and the simulated one, whether that simulation takes the form of a miniature landscape, a map, an exercise, or a purely computational venture like a computer simulation. All these forms rely on and take root in empirical bases: both the questions that these simulations set out to answer and the data used to train and determine the parameters of the model come from measurement, whether experimentally or real-time during disasters. And the problematic of computation—not in concept, but as a raw, quantitative constraint—has informed the practical dimensions of modeling from the beginning periods covered in this book. It is with great historical consequence that fluid dynamics are so computationally demanding to simulate through mathematics. In science fiction, however, no such constraint need exist, so it is within science fiction that we might begin to describe the popular fantasy structure of predictive science in the twenty-first century. In short, now that we have the machines to perform outrageously complicated simulations, what prevailing conceptualizations of prediction drive popular and political discourses? A few examples should serve to anchor the line of inquiry.

Among the best known is Phillip K. Dick's 1956 novella *The Minority Report*, along with its 2002 film adaptation starring Tom Cruise. In it, John Anderton (Cruise) heads the Precrime Unit, which is tasked with preventing crime before it happens. The method relies on the use of clairvoyant "precogs," humans who for some reason receive psychic impressions of the future, and who then transpose that information into a visual medium that police can use to determine the time and place of the crime-to-be. As befits good science fiction thrillers, there are plenty of twists and revelations, not to mention a great deal of Tom Cruise running around and shouting, but the core of the science that animates the premise is resolutely metaphysical: Precrime as a concept and an institution needs precogs and, while it turns out that the emergence of precogs appears to have been the result of gestational drug use, there is nevertheless something magical about the mode of prediction. The science relies entirely on a unique and irreplaceable set of entities, not a method, and not a technology, *per se*. This fantasy of prediction—not as a play of statistical probabilities, but as an ontic object achievable through metaphysical characteristics—is totally foreign to the science of prediction seen through mathematics and statistics.

The magic of computation in the contemporary era provides for a different means than those conceivable in 1956—not precognition, but precomputation. A sequence from the 2020 FX series *Devs* symptomatizes this shift. Or, rather, it provides us with an inroad for thinking about the specific historical articulation—the specific formalization—of the cultural status of the model in an age that ostensibly yearns for an evental break in modernist computational forms (from classical to quantum physics) but which nevertheless grounds its discourse in it.

The series opens with the young Sergei (Karl Glusman) and his partner Lily (Sonoya Mizuno) waking somewhere in the Bay Area, sparring over digital security technologies over breakfast, and then riding a company commuter bus to their shared workplace, tech giant Amaya. Sergei is nervous for a presentation he must give. While each occupies a different professional position (Lily is an engineer, Sergei works in artificial intelligence), the opening minutes establish definitively that their expertise overlaps and, it seems, motivates a portion of their romantic bond. Upon arriving at Amaya, they part ways.

The scene of Sergei's presentation narrates in direct terms the problematic of the contemporary modeling fantasy. He sits, flanked on each side by an ostensible research partner, on one side of a long table as Forest (Nick Offerman), the CEO of Amaya, and his associate Katie (Alison Pill) enter the room and sit on the table's opposite side. Hierarchy having been established, the presentation begins. As an artificial intelligence specialist, Sergei and his team have worked out a computational model that tracks a nematode and its squirming from the instance of its first cell division, also hooking into and mapping its neuronal data over time. Using this information, Sergei and his team have developed a model, visualized by a series of lines and joints, a geometrization of the nematode's body, that performs what must be a vast set of calculations to deploy a simulation whose purpose is to predict the nematode's exact squirming into the future with absolute precision. They overlay the model and a microscopic visual feed of the creature and, after saying "syncing" out loud while typing on their laptops, the model begins to track and conform to the real-time movements of the nematode. Forest, who we learn later possesses an essentially deterministic ontology, anticipates what comes next: he asks them to chart the model ten seconds into the future. The researchers queue the machine, and the jointed model appears to diverge from the real nematode, establishing its own independent, repetitive movements. But as the timer moves closer and then arrives at zero, the model's independent movement is suddenly the nematode's movement, superimposed perfectly, and the model begins to track and conform to the real-time movements of the nematode. A green numerical reading at the bottom of the display indicates the computation's level of accuracy as achieving well above 99 percent.

Forest, impressed with his young employees (though very clearly focused on impressing on Sergei), delightfully proclaims, "So, right in front of my eyes, you just predicted the exact behavior of a living organism, ten seconds into the future." At the very instant that this moment of transference appears about to resolve, the numerical display begins decreasing, rapidly losing its superimposition over the nematode, breaking the algorithmic choreography and ruining the dance until, fading systematically from green to red, it reads 0 percent. Sergei frantically explains that their model can endure for roughly 30 seconds before the computations necessary to process the sheer contingency of the material world—the exponential calculations required to predict even the routine wriggling of a nematode further and further into the future—quiet simply outstrip the capability of the machines performing those equations. This is all to say: the model is correct, and the fantasy of the prediction of the future remains intact. It is merely technological limitations that prevent it from full realization. The remainder of the series unfolds across a philosophical battle between this sort of brute determinism and quantum theories, but, for all the technological nuance it contains, never really escapes the essential conviction that everything set in motion has an end already anticipated; only a good enough computer prevents the uncovering of these truths-to-come. This science of prediction is something entirely different from the one *The Minority Report* presents. It is one based on a relentless faith that every entity, every event, every atom can be accounted for with a godly enough computer. It is a materialism that swallows its own paradoxes whole, is unbothered with the recursivity implied in the whole enterprise—that, in its most banal form, the activity of prediction itself would have been predicted, as would have been the knowledge that one would render predictions about predictions. It's completely ridiculous and entirely coherent, and *Devs* plays with it in a modestly sophisticated manner, even if the plot itself is rather ordinary to the contemporary moment.

However, what sustains these paradoxes is resignation into this determinism: quite simply, the acceptance that the future cannot be changed, that one's attempts to change the future, the capacity and will to want to change the future, and the ideation and action involved in plans to change the future, will have already been predictable and perhaps even predicted. This is one reason for the jumbled mess that many narratives make of complicated time travel scenarios.

Of course, the will to predict is anticipatory, and that anticipation is functional. In plotlines with less investment in determinism as an ontology, prediction functions as a means of control. The HBO series *Westworld*, based on the 1973 Michael Crichton film of the same name, includes as part of the plotline of its third season a quantum computer system called Rehoboam whose primary function is both prediction and orchestration. Rehoboam takes in even the minutest data points of millions upon millions of people, analyzes that data at an astronomical level of sophistication, and charts a course for them. They don't know it, of course, but their lives have been planned out to the detail by Rehoboam, and not knowing—maintaining the feeling of free will and achievement and individuality—is a necessary part of this mass orchestration. To think this process through for even a moment, it must take enormous computational capacity: not only must Rehoboam digest the personal characteristics of every human under its surveillant gaze, it must also forecast how they will interact with their environments, how their environments will respond, how their decision-making will be influenced by the people they meet and the experiences they have, how their very affect will be produced by and within a world of events and disasters and of course other people, all of whom are also being calculated for and given paths of their own, and how unforeseen events might dynamically shift them and require subtle course correction—and much, much more. This utopian theory of harmonious society, a centrally-planned hive of individual nodes, all orchestrated by a supercomputer, but none of whom can know this fact, is of course equally ridiculous and probably downright fascist. But we are not far, in this tale, from either of the other two vignettes. Rehoboam is merely the quantum computational fantasy put into the service of mass discipline, a scaled up version of the Precrime Unit that uses computers instead of precogs and sheds the deterministic materialism of *Devs* in favor of a fairly directly oligarchic structure of social compliance and control.[1]

There is a clear and obvious difference between massively iterative computer simulations used to anticipate weather events and the technologies of prediction seen in popular cultural artifacts like *Minority Report*, *Devs*, and *Westworld*. Repetition and iteration are an integral part of modeling as a method, for its entire conceptualization organizes around the presumption that one can create a framework from which to repeatedly deploy simulations to obtain probabilistic conclusions. It's simply that in the modern era, these simulations are run hundreds, thousands, or millions of times within an intentionally dynamic framework, and then those simulations are collated into their own dataset used to confirm its own set of conclusions. That is, massively iterative simulations are used in the attempt to probabilize simulation itself through the mechanics of its representation much like straightforward statistical modeling probabilizes conclusions through its form. Popular cultural narratives, on the other hand, extend the logic of prediction to the ontological and sometimes even metaphysical, often in the attempt to theorize concepts like time, future, fate, and change.

What these stories do show us, however, is the end-of-the-line fantasy structure of predictive technology—technology, not science—which involves the granulation of data to such a minute level of complexity that probabilistic prediction slips into event-level prediction, which then folds into itself actor- or entity-level prediction. The USACoE, for all its institutional motivations and state-centered prejudices regarding race and capital, pursued genuine lines of scientific inquiry related to the probability, severity, and possible consequences of events that cannot be directly predicted. In the more recent examples, prediction becomes a question of raw computational power and, when the ontology is not directly deterministic, algorithmic adaptability. And there is very little to suggest that this fantasy structure is not authoritarian in nature. Scientists and engineers in the high modernist era sought to unlock the secrets of nature in order to cajole and control it, to scientize and quantify it so that humans could defy it, to build gigantic bridges over great spans or eliminate specific diseases. As Foucault has demonstrated roundly, by extending this logic to the science of human affairs and management, the state could assert the same epistemological imperative in the domains of sociology and economics in the attempt to manage populations and organize its role within a biopolitical regime in which it took responsibility for and power over the guarantee of life. The managerialist approach was (and is) instrumentally defensible in the aggregate, but could not be granularized beyond generalized, historically determined classifications of qualities like psychological traits of criminality or the biology of racial inferiority, and it is only defensible in the aggregate insofar as one does not ask after the conditions of possibility for those aggregations to have become organized the way they can be observed to be. The fantasy of the *individualization* of prediction, however, is the key difference in the algorithmic era.

In late 2023, the headline from a June 2022 Wired article entitled "AI Can Now Predict Crime Before It Happens" began recirculating on social media under the explicit premise that we are heading for a Minority Report-type scenario. The ambiguity of the headline, engineered as most web content is to drive clicks and generate ad revenue, hints at the allure of individualization, almost but not quite implying that these models claim to predict each crime down to actor-level details. Objections to that sort of policing are, to say the least, reasonable and necessary. Inevitably, the story is considerably more complicated: the research study on which the Wired article reports is an attempt to create a predictive heatmap of different sorts of crimes across

Chicago, something the authors claim to have done with 90 percent accuracy. The substance of the actual research, as it turns out, is quite at odds with the Minority Report conclusion. The model's anchor is an interface of geography and event, according to one of its authors: "Our model learns from, and then predicts, event patterns in the urban space, for example, predicting that within a two block radius around the intersection of 67th Street and SW Avenue, there is a high risk of a homicide a week from now. It does not indicate who is going to be the victim or the perpetrator."[2] Perhaps the most consequential takeaway in the research concerns the constitutive, institutional emergence of crime events across space, indicating to the authors what they refer to as a "signature of enforcement bias" as much as a signature of events themselves.[3] The crime heatmaps the authors produce, moreover, allow for a crucial meta-observation that cuts to the core of the article's thesis: that the model actually predicts law enforcement as a primary mechanism and crime only as the effect of that enforcement, its institutional capture, one could say, that inevitably targets certain populations over others and siphons resources from disadvantaged communities into policing.

The ease with which the rich contextualization of crime the article produces slips into the most vulgar conception of targeted policing may not be unreasonable, but it does draw directly on the popular conceptions of prediction science and its ability to atomize. Popular narratives like those explored above are not speaking in terms of return periods or probabilities among populations, i.e., how likely a particular event is to happen within a given population—how likely is a revolt? Instead, the probabilistic function of the individualized prediction returns, full force, to the conceptual question raised by early statistics: what are the chances the model is wrong? Or, inverted for the purposes of state and/or corporate power, what is the likelihood that it is right, and that so-and-so is about to commit a crime or subvert the will of the state?[4] Implicit, too, in this line of inquiry, is the ever-scaling desire for instruments of perfect and unlimited predictive power that can be put to use for any inquiry.

The realm of natural disaster prediction has not been immune from these popular conceptualizations of prediction, nor to the tendency of these public conversations to drive resource allocation to predictive technologies. After the devastation of Hurricane Sandy in 2012, a portion of the public effort to collectively process the event was, as before with Katrina, an assessment of predictive successes and failures. Specifically, news discourse framed part of its analysis through the juxtaposition of the American and European storm track forecasts. The Weather Channel produced a map graphic with the tracks from both models and the actual track of Sandy overlaid.[5] The American model showed Sandy moving north only as far as the latitude of the northern Florida border, and then cutting back northeastward, into the Atlantic. The

European model, in contrast, showed the track hugging the shoreline most of the way up the east coast, and then cutting west into New York state. The actual track was most similar to, and shared the most important defining feature (the cut inland) with the European model. A good deal of the new coverage at the time came to the general conclusion that the European model was scientifically superior, and that that gave some vague indication of lagging U.S. weather technology. If the European model got Sandy right, it must be a better model.[6]

The upshot of these examples is the fixation on *individualization*, the movement from probability to pure prediction, or the tendency of public discourse on predictive science to anchor discussion in the level of the event in its particularity, not merely in its probability or the fact of its occurrence. The eschatology of supercomputational algorithmic modeling is a forecasting of events in their specificity—their timing, the actors involved, the animating motives, and of course the necessary conditions of their realization. When this level of prediction has been achieved, it allows for the total reorganization of events, the complete and unmitigated capacity to prevent things from happening. This scale is, at the present moment, unimaginable except in stories. But the same logic inheres the juxtaposition of the American and European storm models under the rubric of event-level accuracy. Weather scientists and professionals begged the public at the time to understand that experts take numerous models into account before making forecasts, that plenty of other times the American model had been right and the European one wrong, and that fixating on the individual performance of a model in a particular circumstance was neither informative nor scientific. Nevertheless, the idea that predictive modeling, given the best computational conditions (which most often include surveillance as a technology of measurement), can produce individualized forecasts for highly improbable and highly contingent events remains a prevailing feature of the fantasy structure of predictive science across scientific and institutional domains.

What also remains fascinating about the narrative culture around these prediction fantasies, particularly given the naturalism in the very soul of the science, is the cleave between predictive technologies and environmentalism in the crafting of these stories. Still, in cinema, natural disasters are merely what happen, even in the age of anthropocentric climate change. In fact, the fantasy of prediction seems to rely on this rhetorical partitioning: the capacity of a technological means of pure prediction, whether ontological or metaphysical in philosophy, is—more than Kenneth Burke's thesis on the recalcitrance of nature—in fact the resolute *independence* of nature to the science of prediction. It is not incidental, in fact, that the research article on predicting crime cited above must begin with the rejection of a previous prediction method involving modeling techniques derived from seismic disaster modeling—the prior assumption being, of course, that clusters of crime events are directly analogous to the way that earthquakes can trigger smaller afterquakes.[7] For the pure prediction crowd, in spite of the contiguous quantum "observer effect" thesis, wherein the mere observation of a thing effects some sort of ontic change on it, the very conceptualization of pure prediction is founded rhetorically on the assumption of the thing's independence, its determined resignation to carry on as it will and as it must. Only a thoroughly materialist conception of humans-as-nature resolves the paradox, and only instrumentally for the progression of the narrative, not so much philosophically or epistemically. This independence is a crucial condition for individualization and the movement from probability to pure prediction, which I take to be the defining feature of predictive modeling culture in the algorithmic age.

The sometimes subtle and sometimes rather fundamental changes in the rhetoric of the science of prediction are the basis of the thesis that models are memory. Although it is true that predictive science has always been constrained by the limits of technological capacity of one sort or another, this does not mean that rhetorical formulations of prediction have followed inevitably from the progression of the science. The tracing of these forms throughout the course of this book has, on this assumption, attempted to demonstrate their historicization: the sublimation of Black suffering and death in the Jim Crow era, the transposition of bombs into storms and real disasters into preventable ones, and the incorporation of the urge to spectate death in New Orleans (at the same time as the forgotten are disavowed) into the rhetorical form of Katrina's memory objects. As a logical necessity for refusing to read public memory as a progression of objects, I have instead read this shifting of forms as symptomatic of public memory at work: the carrying through of pivotal events in national memory, via rhetoric, within the very form of science itself and in important public debates about that form. Moreover, I have conditioned this reading on the ability to demonstrate that the science of prediction functions as both an indicator and facilitator, so to speak, of the national psychic status of natural disasters—both as a category of event and in their particularity—such that we can refer to science as memory-work.

There were certain necessary limitations to this approach, which I will use as an opportunity to signal future lines of inquiry. First was a heavy focus on state institutional science and rhetoric in the earlier chapters and a more cursory treatment of twenty-first century computerized models in the last chapter. This was both practical and, I think, defensible: the best available literature on twentieth-century disaster modeling routes through the Army Corps of Engineers, and the proliferation of modeling technologies and non-federal institutional modeling capabilities in the contemporary era could justify its own study, with its own careful technical and rhetorical architecture. Rather than a comprehensive review of disaster modeling, which would certainly represent a worthwhile historical endeavor, this book set out to answer a fairly delimited set of questions related to the rhetoric of science, public memory, and the natural, specifically anchored by the national trauma that was Hurricane Katrina. Further work in the rhetoric of science might include, in Foucauldian spirit, more of an archaeology of twenty-first century disaster modeling that cuts across domains like federal research, private and university research, the insurance industry, the financial industry, and foreign policy. Set apart from the task of diagnosing national memory, this sort of work could theorize competing motives and forms; there are lots of reasons, after all, for wanting to know the future.

To that end, this book has also, quite deliberately, taken as given the Beckian, modernist "risk society" approach and the dynamics of the literal marketplace for disasters.[8] I have tended to speak of disaster modeling as a more or less managerial practice, a science of prediction premised on the duty of the state to mitigate against, respond to, and recover after natural disasters (this is in part because to theorize U.S. public memory from the perspective of state and institutional science is at some level to theorize national memory). And those *are* significant tasks that federal and state governments spend enormous amounts of time and money attempting to accomplish, however much I have criticized the rhetorical forms that that science has taken on throughout the twentieth and twenty-first centuries and however much we can point to systemic racism and pro-capitalism in its approach. However, the insurance industry, too, has enormous use of and interest in disaster modeling, and for different explicit purposes. The calculation of the probability and extremity of disaster is central to the calculation of premiums, levels of coverage and, as we've seen recently with insurers abandoning flood insurance availability in Florida, the decision to offer coverage at all. After 1992's Hurricane Andrew, which cost the insurance and reinsurance industry $16.5 billion, a market emergence for catastrophe (CAT) bonds, which essentially use the benchmarks of disaster—a certain dollar amount of losses, or a natural disaster measured to a certain severity—as a threshold for payouts. A year after they were introduced the market for CAT bonds was $2–3 billion.[9] As of 2021, in the wake of devastating wildfires across California and the western United States, it had grown to $12.8 billion.[10] However much a perversion of the institutional conception of disaster management this might seem, it follows the modernist thesis explicitly: an approach to disaster preoccupied with the future and its calculation of risk. Future exploration of the "models as memory" thesis might pose the question explicitly in the context of risk society features, examining the precise ways in which the dynamic malleability of sociological and economic systems can be demonstrated by direct adaptation to the rhetorical formulations of the past's future that risk modeling involves.

Third, while I've made explicit the disavowal of the constitutively social and economic makeup of the category of disaster, and while I've gestured at the irony that upholding this disavowal in the face of anthropogenic climate change presents, a more comprehensive study on the rhetorical dimensions of that disavowal (and that irony) is warranted. The storms are worse, the summers are hotter, and the world just experienced a singular global disaster event in the COVID-19 virus, the differential effects of which country-to-country can only serve as more proof of the critique of the natural. From the interdisciplinary field of disaster studies, important work like Samantha Montano's *Disasterology: Dispatches from the Frontline of the Climate Crisis* provide practicable and sharp analysis of how these events are accelerating and what disaster management institutions can do about it. Scholars studying communication, language, and rhetoric can also continue the important work of analysis, diagnosis, and even advocacy in the face of these collective contradictions: by what rhetorical means does the fantasy of the "natural disaster" function in an age of anthropogenic climate degradation and potential collapse? What are the features of this larger national fantasy that provide the public sensibilities that maintain such a disaster policy regime when, year after year, disaster after disaster, it has proven to be incomprehensibly inadequate to the task of the moment, let alone the future of climate crisis? And not only at the pragmatic level of governance: while the ontological and metaphysical divisions that mark out the boundaries between the human and the natural in the West have long participated in a dubiously reductive form of liberal humanism, the ability to maintain this fantasy must be seen as the result of a rhetorical trick that can be un-troped.

Finally, I have always hoped that this book could serve as a model, so to speak, of a different approach to public memory. Having come to terms with the status of Big Rhetoric, the heart of which was the question of a proper object domain for studying rhetoric, public memory scholars helped carve a new path for reading space and place as rhetorical—for, in a certain sense, "de-objectifying" the study of rhetoric by challenging what can be acceptably called rhetorical. While the study of public memory has taken on many different forms inside the field and out, it therefore does present a certain irony that the vast majority of work produced under its banner today is so strongly overdetermined by the study of space and place, seemingly the proper object domain of public memory. With or without psychoanalysis, which is merely my preferred conceptual lens for reading the rhetorical phenomena I have referred to as displacements, scholars of public memory might begin asking a little more seriously the question of elsewhere when it comes to memory: where are all the non-place sites we collectively work through events of great consequence? It certainly cannot be limited to the objects I've explored here, nor still to science as a mode of inquiry. That stated, the conceptualization of public memory operative in this book is not entirely mutually exclusive with the traditional objects of public memory. It is manifestly true that these objects mediate historical narratives, senses of collective identity, that people are affectively invested in what they are, where they are, and what stories they tell. If this were not so, then I would not have recognized the lack of attention given to the Hurricane Katrina Memorial as being symptomatic of a question to be answered. It is only that, especially for us Freudians, the interface of memory and form is as undetermined as it is inextricable.

* * *

In lots of ways, this remains a project about Hurricane Katrina. While there are now more memory objects devoted to Katrina than when I began, this movement came about only through the decades-long negotiation of the meaning of the storm. On the one hand, it was immediately clear to everyone that Katrina was a human disaster: the levees broke, people were left helpless and hopeless, and the history of racial capitalism in the city that preceded the storm left Black populations vulnerable in a million ways. On the other, it is only through more recent public reckonings about race, vulnerability, and biopolitical arrangements that Katrina emerges as an event more nuanced than merely the story of infrastructural failure: had the levees not failed, things in New Orleans would still not be as great as the mirror-image of pre-disaster tempts us to believe. But the indomitable capacity of the city to anticipate, weather, and move on from disasters of all kinds may be a sort of resilience-through-necessity, and therefore one of limited value for a more comprehensive politics of racial justice, but there may be a model for memory that hums alongside the sequence. At the risk of sounding millenarian, there is something entirely useful about the anticipation of The Big One insofar as it fears the worst. It's quite possible that the narrative fears the worst because it remembers the worst. And at the risk of sounding naïve in the face of racial capitalism and the calculi of neoliberal governance, it invites a more empathetic science that imposes both critical reflexivity on its forms and a decapitalization of its primary motives. What if storm severity was reported not only through the language of return periods, which make disasters feel improbable and without reasonable or recent precedent, but instead through a language of acceptability—would a particular set of consequences ever be acceptable, however improbable? And, first and foremost, what are the lessons we are choosing to learn, and how are we enshrining them in our epistemic and infrastructural practices? To me, there is no more consequential form of public memory than this.

NOTES

1. This is a fascinating choice for the third season, given that the first two seasons of *Westworld*, a show about uncannily lifelike robots wresting control over their inner and outer lives from the humans who built these simulations, could not but hypothesize about things like free will and consciousness. The uncanny is a straightforward psychological and rhetorical device for doing so in the context of simulation.

2. Quoted in: "AI Can Now Predict Crime Before It Happens," *Wired*, July 15, 2022, https://wired.me/technology/ai-can-now-predict-crime-before-it-happens/.

3. Victor Rotaru, Yi Huang, Timmy Li, James Evans, and Ishanu Chattopadhyay, "Event-Level Prediction of Urban Crime Reveals a Signature of Enforcement Bias in US Cities," *Nature Human Behavior* 6 (2022): 1056–68.

4. There is, incidentally, an unfortunate tendency within the surveillance literature in the humanities to take Foucault's adoption of the panopticon a bit too literally: as a series of surveillant technologies (doorbell cameras, "home assistants" like Amazon's Alexa, tracking cookies on the internet, GPS tracking, etc.) that are then used as a means for discipline and control. The panopticon, it should be remembered, was a metaphor for a particular kind of psychology of power, one that founded a subject who acted *as if* they were being watched all the time.

5. See Jacob Feuerstein, "How Hurricane Sandy Sprung Weather Models into the Mainstream," *The Washington Post*, October 29, 2022, https://www.washingtonpost.com/climate-environment/2022/10/29/superstorm-sandy-models-american-european/.

6. Andrew Restuccia, "Sandy Shows Europe Forecasting Lead," *Politico*, November 5, 2012, https://www.politico.com/story/2012/11/sandy-shows-europe-forecasting-lead-083293.

7. Victor Rotaru et al., "Event-Level Prediction," 1056. See also the original reference: G. O. Mohler, M. B. Short, P. J. Brantingham, F. P. Schoenberg, and G. E. Tita, "Self-Exciting Point Process Modeling of a Crime," *Journal of the American Statistical Association* 106, no. 493 (2011): 100–108.

8. Ulrich Beck, *Risk Society: Towards a New Modernity*, trans. Mark Ritter (London: Sage Publications, 1992).

9. Lawrence Richter Quinn, "Predicting Disaster: Is Risk Modeling, Which Claims to Be Able to Predict the Likelihood of Natural Disasters, Science or Wishful Thinking?," *Investment Dealers Digest*, December 21, 1998.

10. Max Reyes, "Catastrophe-Bond Market Hits a Record $12.8 Billion as Extreme Weather Worsens," *Bloomberg*, February 3, 2022, https://www.bloomberg.com/news/articles/2022-02-03/catastrophe-bond-market-hits-record-at-12-8-billion-in-issuance.

Bibliography

"A Mathematical Model for Unsteady-Flow Computations through the Complete Spectrum of Flows on the Lower Ohio River." Accessed January 5, 2024. https://apps.dtic.mil/sti/citations/ADA047242.

Aristotle. *The Basic Works of Aristotle*. Edited by Richard McKeon. Modern Library paperback edition. Modern Library Classics. New York: The Modern Library, 2001.

Badiou, Alain. *Being and Event*. Translated by Oliver Feltham. Paperback edition. Bloomsbury Revelations edition. Bloomsbury Revelations. New York, NY: Bloomsbury Academic, 2007.

———. *The Concept of Model: An Introduction to the Materialist Epistemology of Mathematics*. Edited by Zachary Luke Fraser and Tzuchien Tho. Transmission (Melbourne, Vic.). Melbourne, Australia: re.press, 2007.

Barney, Timothy. *Mapping the Cold War: Cartography and the Framing of America's International Power*. Chapel Hill: University of North Carolina Press, 2015.

Barry, John M. *Rising Tide: The Great Mississippi Flood of 1927 and How It Changed America*. 1st Touchstone ed. New York: Simon & Schuster, 1998.

Baudrillard, Jean. *Simulacra and Simulation*. Translated by Sheila Faria Glaser. Body, in Theory. Ann Arbor: University of Michigan Press, 1994.

Beardsworth, Sara. *Julia Kristeva: Psychoanalysis and Modernity*. SUNY Series in Gender Theory. Albany: State University of New York Press, 2004.

Beck, Ulrich. *Risk Society: Towards a New Modernity*. Theory, Culture & Society (Unnumbered). London: Sage Publications, 1992.

Beer, David. "The Biopolitics of Biometrics: An Interview with Btihaj Ajana." *Theory, Culture & Society* 31, no. 7–8 (December 2014): 329–36.

Bibler, Michael P. "Always the Tragic Jezebel: New Orleans, Katrina, and the Layered Discourse of a Doomed Southern City." *Southern Cultures* 14, no. 2 (2008): 6–27.

Biesecker, Barbara. "No Time for Mourning: The Rhetorical Production of the Melancholic Citizen-Subject in the War on Terror." *Philosophy & Rhetoric* 40, no. 1 (2007): 147–69.

"Blackstone River Flood-Control Project at Woonsocket, Rhode Island: Hydraulic Model Investigation." Accessed January 5, 2024. https://apps.dtic.mil/sti/citations/ADA637651.

Blair, C., M.S. Jeppeson, and E. Pucci. "Public Memorializing in Postmodernity: The Vietnam Veterans Memorial as Prototype." *Quarterly Journal of Speech* 77 (1991): 263–88.

Blair, Carole, V. William Balthrop, and Neil Michel. "The Arguments of the Tombs of the Unknown: Relationality and National Legitimation." *Argumentation* 25, no. 4 (November 2011): 449–68.

Brock, Kevin. *Rhetorical Code Studies: Discovering Arguments in and around Code*. University of Michigan Press, 2019.

Butler, Judith. *The Psychic Life of Power: Theories in Subjection*. Stanford, Calif.: Stanford University Press, 1997.

Cagle, Lauren E. "Surveilling Strangers: The Disciplinary Biopower of Digital Genre Assemblages." *Computers and Composition* 52 (June 2019): 67–78.

Camiller, Patrick, trans. *War and Cinema: The Logistics of Perception*. Radical Thinkers. London: Verso, 1989.

Cheramie, Kristi. "The Scale of Nature: Modeling the Mississippi River." *Places Journal*, March 21, 2011.

Chow, Ven Te, ed. *Handbook of Applied Hydrology*. 2nd ed. New York: McGraw-Hill Education, 1964.

Condit, Celeste M., John Lynch, and Emily Winderman. "Recent Rhetorical Studies in Public Understanding of Science: Multiple Purposes and Strengths." *Public Understanding of Science* 21, no. 4 (May 1, 2012): 386–400.

Copjec, Joan. *Read My Desire: Lacan against the Historicists*. Radical Thinkers. Cambridge, Mass.: MIT Press, 2015.

Cotton, Gordon A. *A History of the Waterways Experiment Station, 1929–1979*. Vicksburg, Mississippi: U.S. Army Corps of Engineers, Waterway Experiment Station, 1979. https://archive.org/details/historyofwaterwa00vick.

"Department of State Washington File: Transcript: Defense Department Briefing, February 12." Accessed January 5, 2024. https://usinfo.org/wf-archive/2002/020212/epf202.htm.

Derrida, Jacques. *Dissemination*. Translated by Barbara Johnson. Chicago: University Press, 1981.

Derrida, Jacques, Catherine Porter, and Philip Lewis. "No Apocalypse, Not Now (Full Speed Ahead, Seven Missiles, Seven Missives)." *Diacritics* 14, no. 2 (1984): 20–31.

Dhanak, Manhar R., and Nikolaos I. Xiros, eds. *Springer Handbook of Ocean Engineering*. Cham: Springer International Publishing, 2016.

Dickinson, Greg, Carole Blair, and Brian L. Ott. *Places of Public Memory: The Rhetoric of Museums and Memorials*. Rhetoric, Culture, and Social Critique. Tuscaloosa: University of Alabama Press, 2010.

DuBois, W.E.B. "Postscript." *The Crisis*, no. 34 (November 1927).

Dunn, Thomas R. "'The Quare in the Square': Queer Memory, Sensibilities, and Oscar Wilde." *Quarterly Journal of Speech* 100, no. 2 (April 3, 2014): 213–40.

Dyson, Michael Eric. *Come Hell or High Water: Hurricane Katrina and the Color of Disaster*. 1st ed. New York: Civitas Books, 2007.

Ekeland, I. *Mathematics and the Unexpected.* Chicago: University of Chicago Press, 1988.

Elliott, Dabney Otey. *The Improvement of the Lower Mississippi River for Flood Control and Navigation.* St. Louis, Mo.: M.R.C. May 1, 1932.

Farish, Matthew. "Disaster and Decentralization: American Cities and the Cold War." *Cultural Geographies* 10, no. 2 (2003): 125–48.

Fatherree, Ben H. *The First 75 Years: History of Hydraulics Engineering at the Waterways Experiment Station.* Vicksburg, MS: US Army Corps of Engineers, Engineer Research and Development Center, 2004.

Fields, Barbara Jeanne. "Slavery, Race and Ideology in the United States of America." *New Left Review*, 181, no. 1 (1990): 95–118.

Fisher, R. A., and Edward John Russell. "On the Mathematical Foundations of Theoretical Statistics." *Philosophical Transactions of the Royal Society of London. Series A, Containing Papers of a Mathematical or Physical Character* 222, no. 594–604 (January 1997): 309–68.

Foster, J. E. *History and Description of the Mississippi Basin Model. Mississippi Basin Model. Report 1-6.* Waterways Experiment Station (U.S.). Vicksburg, Miss.: U.S. Army Engineer Waterways Experiment Station, 1971.

Foucault, Michel. *The Archaeology of Knowledge; and the Discourse on Language*. Translated by Alan Sheridan. Vintage books ed. New York: Vintage Books, 2010.

———. *The Birth of Biopolitics: Lectures at the Collège de France, 1978–79*. Edited by Michel Senellart and Graham Burchell. Michel Foucault: Lectures at the Collège de France. New York: Picador, 2004.

———. *The Order of Things: An Archaeology of Human Sciences*. New York: Vintage Books, 1970.

Fowle, Barry W. *Builders and Fighters U.S. Army Engineers in World War II.* Chicago: Barakaldo Books, 2021.

Freeman, John R. "The Need of a National Hydraulic Laboratory for the Solution of River Problems." *Transactions of the American Society of Civil Engineers* 87, no. 1 (January 1924): 1033–97.

Freud, Sigmund. *Introductory Lectures on Psycho-Analysis*. Edited by James Strachey. New York: W. W. Norton, 1966.

———. *Moses and Monotheism*. Translated by Katherine Jones. New York: Vintage Books, 1939.

———. *Standard Edition of the Complete Psychological Works of Sigmund Freud, Volume XXIII (1937–1939): Moses and Monotheism, An Outline of Psycho-Analysis and Other Works*. Translated by J. Strachey. London: Hogarth Press, 1964.

———. *The Ego and the Id.* New York: W. W. Norton, 1960.

———. *The Interpretation of Dreams*. Edited by James Strachey. New York: Basic Books, a Member of the Perseus Books Group, 2010.

Fuller, Steve, and James H. Collier. *Philosophy, Rhetoric, and the End of Knowledge: A New Beginning for Science and Technology Studies.* 2nd ed. Mahwah, N.J.: Lawrence Erlbaum Associates, 2004.

Gotham, Kevin Fox. "(Re)Branding the Big Easy: Tourism Rebuilding in Post-Katrina New Orleans." *Urban Affairs Review* 42, no. 6 (July 1, 2007): 823–50.

———. "Touristic Disaster: Spectacle and Recovery in Post-Katrina New Orleans." *Geoforum* 86 (November 2017): 127–35.

Graham, Howard E., and Dwight E. Nunn. "Meteorological Considerations Pertinent to Standard Project Hurricane, Atlantic and Gulf Coasts of the United States." National Hurricane Research Project, 1959.

Grossman, Jeremy R. "Take 'Em Down: Rhetorical Temporality and Critical Regionalism in the Struggle to Remove Confederate Monuments in New Orleans." In *Reconstructing Southern Rhetoric*, by Jeremy R. Grossman, 126–43. Jackson: University Press of Mississippi, 2021.

———. "The Catastrophe to Come: Prefiguring Hurricane Katrina's Public Memory through the Anxious Melancholic Rhetoric of 'the Big One.'" *Rhetoric and Public Affairs* 23, no. 3 (2020): 417–54.

———. "Dreaming of Disaster: Displacements of Public Memory and Hurricane Katrina." University of Georgia, 2018. http://purl.galileo.usg.edu/uga_etd/grossman_jeremy_r_201805_phd.

Gunn, Joshua. "Hystericizing Huey: Emotional Appeals, Desire, and the Psychodynamics of Demagoguery." *Western Journal of Communication* 71, no. 1 (2007): 1–27.

Hallsby, Atilla. "Imagine There's No President: The Rhetorical Secret and the Exposure of Valerie Plame." *Quarterly Journal of Speech* 101, no. 2 (2015): 354–78.

Harding, Sandra G. *Whose Science? Whose Knowledge?: Thinking from Women's Lives*. Science Question in Feminism. Ithaca, N.Y.: Cornell University Press, 1991.

Hariman, Robert, and John Louis Lucaites. *The Public Image: Photography and Civic Spectatorship*. Chicago: The University of Chicago Press, 2016.

Hartnell, Anna. "Katrina Tourism and a Tale of Two Cities: Visualizing Race and Class in New Orleans." *American Quarterly* 61 (2009): 723.

History of the Waterways Experiment Station. Waterways Experiment Station, 1968. http://cdm16021.contentdm.oclc.org/cdm/ref/collection/p16021coll4/id/84.

Horowitz, Andy. *Katrina: A History, 1915–2015*. Cambridge, MA: Harvard University Press, 2020.

Huet, Marie-Hélène. *The Culture of Disaster*. Chicago, IL: University of Chicago Press, 2012.

"Hurricane Pam Exercise Concludes." FEMA, July 23, 2004. https://biotech.law.lsu.edu/blaw/FEMA/FEMA_Hurricane_Pam.pdf.

Hutton, Patrick H. *History as an Art of Memory*. Lebanon: University Press of New England, 1993.

Hydraulic Models as an Aid to the Development of Design Criteria. Waterways Experiment Station (U.S.). Vicksburg, Miss.: Waterways Experiment Station, 1951.

Iklé, Fred Charles. *The Social Impact of Bomb Destruction*. 1st ed. Norman: University of Oklahoma Press, 1958.

Iliadis, Andrew. "Algorithms, Ontology, and Social Progress." *Global Media and Communication* 14, no. 2 (2018): 219–30.

Interlying Areas along Coastal Louisiana in the Vicinity of Houma: Letter from the Secretary of the Army Transmitting a Letter from the Chief of Engineers, Department of the Army, Dated March 6, 1967, Submitting a Report ... on an Interim Hurrican

Study of a Portion of Coastal Louisiana in the Vicinity of Houma, Authorized by Public Law 71, 84th Congress, Approved June 15, 1955. Washington, D.C.: U.S. G.P.O., 1967. http://catalog.hathitrust.org/api/volumes/oclc/85266004.html.

Investigation of the Performance of the New Orleans Flood Protection Systems in Hurricane Katrina on August 29, 2005. Berkeley, CA: Independent Levee Investigation Team, 2006. http://www.ce.berkeley.edu/projects/neworleans/draft.htm.

Jadwin, Edgar. "The Plan for Flood Control of the Mississippi River in Its Alluvial Valley. Presented to the Secretary of War and by Him to Congress." *The Annals of the American Academy of Political and Social Science* 135 (1928): 35–44.

Johnson, Cedric, ed. *The Neoliberal Deluge: Hurricane Katrina, Late Capitalism, and the Remaking of New Orleans*. Minneapolis: University of Minnesota Press, 2011.

Jones, Steve. "People, Things, Memory and Human-Machine Communication." *International Journal of Media and Cultural Politics* 10 (2014): 245.

Jordan, John M. *Machine-Age Ideology: Social Engineering and American Liberalism, 1911–1939*. Chapel Hill: University of North Carolina Press, 1994.

Kelman, Ari. *A River and Its City: The Nature of Landscape in New Orleans: With a New Preface*. Berkeley: University of California Press, 2006.

King, Claire Sisco. "Rogue Waves, Remakes, and Resurrections: Allegorical Displacement and Screen Memory in 'Poseidon.'" *Quarterly Journal of Speech* 94 (November 2008): 430.

Klein, Naomi. *The Shock Doctrine: The Rise of Disaster Capitalism*. 1st ed. New York: Metropolitan Books/Henry Holt, 2007.

Kristeva, Julia. *Pulsions du Temps*. Paris: Fayard, 2013.

———. *Revolution in Poetic Language*. Translated by Margaret Waller Introduction by Leon S. Roudiez. New York: Columbia University Press, 1974.

Lacan, Jacques. *Écrits: The First Complete Edition in English*. Translated by Bruce Fink, Héloïse Fink, and Russell Grigg. New York: W.W. Norton & Co., 2006.

Laclau, Ernesto. *On Populist Reason*. London: Verso, 2005.

Lapp, R. E. "Atomic Bomb Explosions — Effects on an American City." *Bulletin of the Atomic Scientists* 4, no. 2 (February 1, 1948): 49–54.

Larabee, Ann E. "Ground Zero: The City, the Bomb and the End of History." *Canadian Review of American Studies* 22, no. 2 (September 1, 1991): 263–80.

Layton, Edwin T. "Escape from the Jail of Shape; Dimensionality and Engineering Science." In *Technological Development and Science in the Industrial Age: New Perspectives on the Science-Technology Relationship*, edited by Peter Kroes and Martijn Bakker, 35–68. Boston Studies in the Philosophy of Science, 1992.

Leroy, Justin, and Destin Jenkins, eds. *Histories of Racial Capitalism*. New York: Columbia University Press, 2021.

Lessl, Thomas M. *Rhetorical Darwinism: Religion, Evolution, and the Scientific Identity*. Studies in Rhetoric and Religion. Waco, Tex.: Baylor University Press, 2012.

Lévi-Strauss, Claude: *Structural Anthropology: V.1*. New edition. London: Penguin 1993.

Lichtman, Sarah A. "Do-It-Yourself Security: Safety, Gender, and the Home Fallout Shelter in Cold War America." *Journal of Design History* 19 (2006): 39–55.

Llobera, Joseph R. "Halbwachs, Nora and 'History' versus 'Collective Memory': A Research Note." *Durkheimian Studies* 1 (1995): 35–44.

Lundberg, Christian O. *Lacan in Public: Psychoanalysis and the Science of Rhetoric*. Rhetoric, Culture, and Social Critique. Tuscaloosa: University of Alabama Press, 2012.

Mahony, P. "Toward the Understanding of Translation in Psychoanalysis." *Journal of the American Psychoanalytic Association* 28 (1980): 461.

McEnaney, Laura. *Civil Defense Begins at Home: Militarization Meets Everyday Life in the Fifties*. Princeton, NJ: Princeton University Press, 2000.

Mizelle, Richard M. *Backwater Blues: The Mississippi Flood of 1927 in the African American Imagination*. Minneapolis: University of Minnesota Press, 2014.

Mohler, G. O., M. B. Short, P. J. Brantingham, F. P. Schoenberg, and G. E. Tita. "Self-Exciting Point Process Modeling of Crime." *Journal of the American Statistical Association* 106, no. 493 (2011): 100–108.

Monteyne, David. *Fallout Shelter: Designing for Civil Defense in the Cold War*. Architecture, Landscape, and American Culture Series. Minneapolis: University of Minnesota Press, 2011.

Morris, Christopher. *The Big Muddy: An Environmental History of the Mississippi and Its Peoples, from Hernando de Soto to Hurricane Katrina*. Oxford: Oxford University Press, 2012.

Morrison, Margaret. *Reconstructing Reality: Models, Mathematics, and Simulations*. Oxford Studies in Philosophy of Science. Oxford: Oxford University Press, 2015.

Mumford, Lewis. *Technics and Civilization*. Chicago, IL: The University of Chicago Press, 2010.

Nadesan, Majia Holmer. "Hurricane Katrina: Governmentality, Risk, and Responsibility." *Controversia* 6, no. 1 (2008): 67–91.

Nelson, S. C., J.-H. Yu, and L. Ceccarelli. "How Metaphors About the Genome Constrain CRISPR Metaphors: Separating the 'Text' From Its 'Editor.'" *The American Journal of Bioethics* 15, no. 12 (2015): 60–62.

Nora, Pierre. *Realms of Memory: Rethinking the French Past*. Translated by Lawrence D. Kritzman. European Perspectives. New York: Columbia University Press, 1996.

Parrish, Susan Scott. *The Flood Year 1927: A Cultural History*. Princeton: Princeton University Press, 2017.

Pezzullo, Phaedra C. "'This Is the Only Tour That Sells': Tourism, Disaster, and National Identity in New Orleans." *Journal of Tourism and Cultural Change* 7, no. 2 (2009): 99–114.

———. "Tourists and/as Disasters: Rebuilding, Remembering, and Responsibility in New Orleans." *Tourist Studies* 9, no. 1 (2009): 23–41.

———. *Toxic Tourism: Rhetorics of Pollution, Travel, and Environmental Justice*. Tuscaloosa: University of Alabama Press, 2007.

Phillips, Kendall R. *Framing Public Memory*. Rhetoric, Culture, and Social Critique. Tuscaloosa: University of Alabama Press, 2004.

———. "The Failure of Memory: Reflections on Rhetoric and Public Remembrance." *Western Journal of Communication* 74 (2010): 208.

Rayleigh, Lord. "The Principle of Similitude." *Nature*, no. 95 (1915): 66–68.

Resio, Donald T., Jennifer L. Irish, Joannes J. Westerink, and Nancy J. Powell. "The Effect of Uncertainty on Estimates of Hurricane Surge Hazards." *Natural Hazards* 66, no. 3 (2013): 1443–59.

Reuss, Martin. "The Art of Scientific Precision: River Research in the United States Army Corps of Engineers to 1945." *Technology and Culture* 40 (1999): 292–323.

"Revised Standard Project Hurricane Criteria for the Atlantic and Gulf Coasts of the United States." U.S. Department of Commerce, National Oceanic and Atmospheric Administration, 1972.

Rheinberger, Hans-Jörg. "Transpositions: From Traces through Data to Models and Simulations." In Transpositions: Aesthetico-Epistemic Operators in Artistic Research, edited by Michael Schwab, 215–24. Leuven, Belgium: Leuven University Press, 2018.

Ricoeur, Paul. *Memory, History, Forgetting*. Translated by Kathleen Blamey and David Pellauer. Chicago, IL: University of Chicago Press, 2006.

Rose, Kenneth D. *One Nation Underground: The Fallout Shelter in American Culture*. New York: NYU Press, 2001.

Rotaru, Victor, Yi Huang, Timmy Li, James Evans, and Ishanu Chattopadhyay. "Event-Level Prediction of Urban Crime Reveals a Signature of Enforcement Bias in US Cities." *Nature Human Behaviour* 6, no. 8 (2022): 1056–68.

Roundtree, Aimee Kendall. *Computer Simulation, Rhetoric, and the Scientific Imagination: How Virtual Evidence Shapes Science in the Making and in the News*. Lanham, MD: Lexington Books, 2014.

Rubenstein, Diane. *This Is Not a President: Sense, Nonsense, and the American Political Imaginary*. New York: NYU Press, 2008.

Salvaggio, John E. *New Orleans' Charity Hospital: A Story of Physicians, Politics, and Poverty*. Baton Rouge: Louisiana State University Press, 1992.

Shallat, Todd. *Structures in the Stream: Water, Science, and the Rise of the U.S. Army Corps of Engineers*. 1st ed. American Studies Series. Austin: University of Texas Press, 1994.

Smith, Mark M. *Camille, 1969: Histories of a Hurricane*. Mercer University Lamar Memorial Lectures. Athens: University of Georgia Press, 2011.

"Southeast Louisiana Catastrophic Hurricane Functional Plan: Draft." Baton Rouge, LA: Innovative Emergency Management, 2004.

Standard Project Flood Determinations. Civil Works Engineer Bulletin 52-8. Department of the Army, Office of the Chief of Engineers, 1952.

Stow, Simon. "Do You Know What It Means to Miss New Orleans? George Bush, the Jazz Funeral, and the Politics of Memory." *Theory & Event* 11, no. 1 (2008).

Sturken, Marita. *Tangled Memories: The Vietnam War, the AIDS Epidemic, and the Politics of Remembering*. Berkeley: University of California Press, 1997.

The Mississippi River Flood-Control Model. Vol. 37. The Experiment Station Hydraulics Bulletin, 1942.

The US Government Printing Office. "The United States Strategic Bombing Survey: The Effects of Atomic Bombs on Hiroshima and Nagasaki, June 30, 1946." , 1946.

Thomas, Lynnell L. *Desire and Disaster in New Orleans: Tourism, Race, and Historical Memory*. Durham: Duke University Press, 2014.

Tuggle, Lindsay. "Encrypting Katrina: Traumatic Inscription and the Architecture of Amnesia." *InVisible Culture*, March 20, 2011.

US Department of Commerce, NOAA. "Mississippi River Flood History 1543–Present." NOAA's National Weather Service. Accessed January 5, 2024. https://www.weather.gov/lix/ms_flood_history.

Vivian, Bradford. *Public Forgetting: The Rhetoric and Politics of Beginning Again*. University Park: The Pennsylvania State University Press, 2010.

Vogel, Herbert D. "Origin of the Waterways Station." *The Military Engineer*, no. 352 (1961): 132–35.

———. *Paper Y: Experiment to Determine the Limit of Backwater Influence in the Illinois River*. Waterways Experiment Station (U.S.). Vicksburg, Miss.: Waterways Experiment Station, 1931.

Willson, Michele. "Raising the Ideal Child? Algorithms, Quantification and Prediction." *Media, Culture & Society* 41, no. 5 (2019): 620–36.

Wright, Richard. "Silt." *The New Masses*, no. August 24, 1937.

Wright, Thomas. "Scale Models, Similitude and Dimensions: Aspects of Mid-Nineteenth-Century Engineering Science." *Annals of Science* 49 (1992): 233–54.

Žižek, Slavoj. *Enjoy Your Symptom!: Jacques Lacan in Hollywood and Out*. New York: Routledge, 2008.

———. "Joe Public v the Volcano." *New Statesman*, April 29, 2010. https://zizek.uk/2010/04/29/joe-public-v-the-volcano/.

———. *The Parallax View*. Short Circuits. Cambridge, Mass.: MIT Press, 2006.

———. *The Sublime Object of Ideology*. New ed. Essential Žižek. London: Verso, 1989.

Index

About the Author

Jeremy R. Grossman runs a digital marketing agency in Washington, DC. He holds a PhD in Communication Studies and currently teaches classes at the University of Maryland.